Event History Analysis
In Demography

Event History Analysis In Demography

Daniel Courgeau
and
Éva Lelièvre

CLARENDON PRESS · OXFORD
1992

Oxford University Press, Walton Street, Oxford OX2 6DP
Oxford New York Toronto
Delhi Bombay Calcutta Madras Karachi
Petaling Jaya Singapore Hong Kong Tokyo
Nairobi Dar es Salaam Cape Town
Melbourne Auckland
and associated companies in
Berlin Ibadan

Oxford is a trade mark of Oxford University Press

Published in the United States
by Oxford University Press, New York

© Institut National D'Études Démographiques 1992

All rights reserved. No part of this publication may be reproduced, stored in a retrieval system, or transmitted, in any form or by any means, electronic, mechanical, photocopying, recording, or otherwise, without the prior permission of Oxford University Press

British Library Cataloguing in Publication Data
Data available

Library of Congress Cataloging in Publication Data
Courgeau, Daniel.
Event history analysis in demography / Daniel Courgeau and Éva Lelièvre
p. cm.
Translated from the French.
Includes bibliographical references and index.
1. Event history analysis. 2. Longitudinal method. 3. Regression analysis.
I. Lelièvre, Eva. II. Title.
H61.C6793 1992 304.6′01′519536—dc20 92-2511
ISBN 0-19-828738-0

Typeset by Pure Tech Corporation, Pondicherry, India
Printed in Great Britain by
Bookcraft (Bath) Ltd, Midsomer
Norton, Avon

Acknowledgments

First of all, we wish to thank the participants to the 'Séminaire de formation aux méthodes d'analyse biographique' (Training seminar on the methods of event history analysis) (INED, 24–28 August 1987), who firmly recommended that we write up the theoretical and practical presentation we had done: Philippe Antoine (ORSTOM), Didier Blanchet (INED), Catherine Bonvalet (INED), Marco Bottai (Istituto di Statistica, Pisa, Italy), Michel Bozon (INED), Arnaud Bringé (INED), Christiane Delbès (Fondation Nationale de Gérontologie), Jean-Marie Firdion (INED), Fatima Juarez (El Colegio de Mexico, Mexico), Monique Lefebvre (INED), Henri Leridon (INED), Gilles Pison (INED), Denise Pumain (INED), Benoit Riandey (INED), Eric Sammartino (INED), Laurent Toulemon (INED), and Bernard Zarca (CREDOC).

We should also like to thank the members of the jury of the thesis that Éva Lelièvre presented at the Ecole des Hautes Etudes en Sciences Sociales on 23 March 1988, on the 'Méthodes mathématiques et statistiques pour l'analyse d'histoires de vie': Marc Barbut (EHESS), Bui Trong Lieu (Université de Paris V), Hervé Le Bras (INED), and Denise Pumain (INED).

We are also grateful to Jan Hoem (University of Stockolm, Sweden) for the many exchanges, both written and oral, that we had with him on the topic of this book.

Finally, we thank Françoise Milan for typing the manuscript, Nicole Berthoux for producing the figures and graphs, Hella Courgeau for publishing this textbook in French, and MEDIANE (Paris) for translating it into English.

D.C.
E.L.

Contents

Introduction   1

    I. Extending the Scope of Longitudinal Analysis   7

1. Observation of Event Histories   9

1.1. Various types of surveys   11
1.2. The 'Triple Biographie' survey   13
1.3. The problem of weight   16
1.4. Incomplete and erroneous survey observation   18
1.5. The problem of censoring   26
1.6. Conclusion   28

2. Formalization of the Analysis   29

2.1. Analysis of a homogeneous cohort experiencing a single event   30
2.2. Analysis of a heterogeneous cohort and of the interaction between phenomena   36
2.3. Towards a more exhaustive analysis of human behaviour patterns   45
2.4. Conclusion   48

3. Methods of Estimation using Censored Observations   50

3.1. Censoring problems   50
3.2. Right-censoring   52
3.3. Left-censoring   61
3.4. Conclusion   66

4. Study of a Single Event   68

4.1. Single sample: a single event   68
4.2. Single sample: competing risks   72
4.3. Multiple samples: comparative tests   75
4.4. Conclusion   81

5. Reciprocal Study of Interactions between Two Events — 82

5.1. Conception of the analysis — 82
5.2. The bivariate case — 87
5.3. Practical analysis — 93
5.4. Conclusion — 98

6. Extending to More Complex Situations — 99

6.1. Presentation and limits of the practical application — 99
6.2. Interactions between three events: two study cases — 101
6.3. Interactions between two renewable processes — 104
6.4. Conclusion — 106

II. Extending the Scope of Regression Models — 107

7. Statistical Formalization of Parametric Analysis — 109

7.1. Some useful parametric models in demography — 109
7.2. Regression models — 135
7.3. Conclusion — 144

8. Methods of Estimation of Parametric Models — 145

8.1. Computation of the likelihood in the presence of censoring — 146
8.2. Estimating the parameters and testing their value — 148
8.3. Estimation of the parameters: examples — 149
8.4. Comparison of parametric models — 173
8.5. Conclusion — 177

9. Methods of Semi-parametric Analysis — 180

9.1. From parametric regressions to semi-parametric proportional hazard models — 180
9.2. Methods of estimation — 182
9.3. The Newton–Raphson algorithm — 187
9.4. The choice of a model for the analysis of interactions — 188
9.5. Some applications — 189
9.6. Conclusion — 193

10. Conclusion — 195

10.1. Analysis of interactions between phenomena — 195

10.2. Dealing with heterogeneity in populations	200
10.3. New lines of research	204
Appendix	206
Bibliography	210
Index	221

Introduction

Up to the present, longitudinal analysis has basically been developed through an approach that takes each demographic phenomenon into consideration separately. Its main objective has been to isolate each phenomenon in its 'pure state'. In particular, it was considered necessary to separate the effects of each demographic variable from those of the others, such as mortality and migrations. To this end, a certain number of hypotheses were required, but they could not be tested against existing sources (Henry, 1959, 1966).

Longitudinal analysis was developed on the basis of aggregate data such as registration statistics, or, whenever possible, data from population registers. Even though these sources allow each event to be dealt with separately, since they eliminate disturbing effects, they hardly make it possible to analyse interactions between different phenomena. This explains why, in traditional demographic manuals, we find isolated phenomena in their 'pure state' presented as the subject matter of separate chapters: nuptiality, fertility, mortality, moves and migrations (Pressat, 1961; Henry, 1972).

None the less, certain demographers have pointed out the usefulness of analysing interactions between demographic phenomena. Pressat (1966) emphasized that 'the search for correlations between demographic phenomena, even though this domain remains unexplored, should enable us to deepen our knowledge considerably'. He did not, however, give any indication of what method to follow. Similarly, Henry (1972), in his analysis of nuptiality, stated that, 'in the case of out-migrants, one might be tempted to substitute their nuptiality abroad for that which they would have experienced had they stayed in their original country; yet nuptiality in a foreign country depends on conditions that may differ considerably'. This is a definite recognition of the interaction between the two phenomena and the change in an individual's nuptial behaviour following his out-migration. As data were lacking, however, Henry did not pursue this analysis any further.

Another important problem is that of heterogeneity. This point was also touched on by Henry (1959). Although, 'in the case of a homogeneous cohort, the statistical history of the component individuals is the same as the cohort's statistical history', this result no longer holds true when a heterogeneous cohort is being studied. Thus, for example, in the simplest of cases, where two sub-populations each have a constant though different probability of occurrence of the studied event over the period, the overall population will no longer have a constant rate, taken as an average of the two sub-populations' rates. It may well be the case at the outset, but with time the sub-population with the highest hazard rate will be eliminated from the population at risk by a selection process. This means that after a while the hazard rate of the observed population will converge with that of the sub-population with the lowest hazard rate. This heterogeneity may, of course, be of a much more complex nature and would need to be studied further.

A precise knowledge of the practical implications of heterogeneity in human groups would necessitate further differential demographic research into the individual's physical and psychological characteristics, in order to study both the dispersion and correlation of the intra-group demographic indices which have so far been studied in a rather cursory fashion. (Henry, 1959)

As long as demographers use statistics such as those published in registration records or population registers, they have no way of dealing with the two basic problems: the analysis of interactions between demographic phenomena, and the analysis of heterogeneity in human groups. Other sources must now be used in order to observe a group of individuals over their entire lifetime, or at least part of it, as well as to collect a greater number of characteristics for each respondent.

It can thus be seen that the unit of analysis will no longer be the event (death, marriage, birth, migration, etc.): instead, each individual biography will be viewed as a more complex process. The question is no longer one of trying to isolate each phenomenon in its 'pure state': on the contrary, we must try and see how one event in an individual's existence can influence his life-course, and how certain characteristics can induce an individual to adopt behaviour patterns that are different from those of another individual.

This change of view leads us to reformulate the fundamental notions of demographic analysis in terms of complex stochastic processes. Let us take a more detailed look at how this is to be done.

Demographic processes do not occur in an abstract space–time, but originate within a given social structure. Someone born into a Lobi tribe in the early twentieth century will have quite a different biography from someone born into rural France of the same period, or someone born into present-day urban France. In each of these social structures, however, it is possible to distinguish relational systems that have developed to a greater or lesser extent depending on the group or the society concerned: 'family, economic, political, religious, educational, associational and informal systems' (Kimball and Pearsall, 1954). There is, of course, nothing to prevent new types of relational systems from appearing in the future. Our approach does not consider a society as a closed entity, but rather as one in constant evolution.

Each member of a given society is simultaneously involved in the various systems. For example, someone living in present-day France may be part of the family system as a member of an unmarried couple and father of a child; in the economic system as an engineer in the car industry; in the political system as a town councillor; in the religious system as a non-practising Catholic; in the educational system as a recipient of professional training; in the associational system as an amateur footballer; and, finally, in the informal system through his occasional attendance at parent–teacher meetings to solve his child's educational problems.

It is this interaction between the different types of involvement that creates a space and time specific to each situation. The geographical or occupational mobility of a single person may be much more frequent and may occur over greater distances than that of a married person, especially if the latter has one or more children. The married person is naturally tied to his place of residence and work by constraints related to his spouse's work-place, the location of his children's school, etc.

Event history analysis will thus attempt to place these changes in the time and space of an individual's life, in his social context. The point is to see how an event of a family, economic or other nature experienced by the individual will change the probability of other events happening to him over his lifetime. We shall, for instance, try to discover how his marriage can influence his professional career, his spatial mobility and other occurrences, such as the birth of a child or a break with his original family ties.

Here we are directly concerned with the analysis of interactions between demographic phenomena, the utility of which we have already pointed out. This method of analysis has its place in the study of event histories.

Similarly, when trying to understand an individual's behaviour, one must take into account his social origins and his entire past history. In this case we are supposing that behaviour patterns are not innate but rather that they can change over an individual's lifetime as a result of what he experiences and acquires with time. Thus, two individuals from the same social background, but who have taken entirely different paths in life, can have attitudes to marriage, forming a family, career, etc., that diverge increasingly as time goes on.

We, thus, arrive at a method of analysis of population heterogeneity which uses a dynamic rather than static approach and which, accordingly, has its place in the study of individual event histories.

It is important to note that this analysis of population heterogeneity is not deterministic, but basically probabilistic.

Consequently, a large number of individuals who find themselves in the same conflictual situation at the outset will have different probabilities of finding solutions to the situation before a given date. Some will never find solutions; others may invent for themselves an entirely new pattern of behaviour which can provide a better solution to their conflictual situation. We are, thus, allowing the individual a margin of freedom which may lead to entirely new situations. Such a margin of freedom is of course essential, as no chain of events is predestined, but evolves with the course of time.

On this point, we are very close to Prigogine and Stengers (1988), who recognize that 'the event creates a difference between the past and the future ... It is the intelligible outcome of a past, from which, however, it could not be deduced. It opens out an historic future where the insignificance or meaning of its consequences will be decided.' Here, one finds that same margin of freedom which can lead to entirely new situations.

After this informal introduction to event history analysis, can we now proceed to describe it in more formal terms?

When an individual is born, his life can follow a wide variety of paths. These different life-courses, however, are far from being equally probable. An individual's event history can therefore be defined as the result of a complex stochastic process, which develops over time, yet is situated within given historical, geographical, economic, and social conditions.

Introduction

Let Ω_θ represent the set of event histories or partial event histories which can be observed up to a time θ. As already stated, our observation must be limited to the past, since the future brings into play new situations that cannot be deduced from the past. For example, the career of garage-owner could not have been envisaged before the appearance of motor cars, and in the near future genetic discoveries may stretch the span of human life to 150 years or more. The analysis carried out at time t only takes into account past behaviour patterns and projects them into the future without introducing elements affecting their evolution. This evolution, however, takes place at a slow enough rhythm to enable the analysis of the past to enlighten us as to the probability of various events occurring in the near future.

Let ω_n denote an entirely observed event history, where n precedes θ, and ω_θ an event history observed up to the moment θ, which is not finished. It can be said that the events in either of these histories are variables defined in the general space of Ω_θ. For example, the age at which one of these individuals marries in an *application* of Ω_θ on $[0, +\infty]$.

Now, let us give Ω_θ with a sigma-algebra[1] $\mathcal{B}_\theta$ of events that are to be analysed within a given population, together with a probability measure P_θ of $\mathcal{B}_\theta$ on $[0,1]$, which assigns the different probabilities to the different events occurring within the observed population. In this case, $(\Omega_\theta, \mathcal{B}_\theta, P_\theta)$ does indeed define a probability space.

A random moment T will thus be a function of time on the Ω_θ probability space, which can in this case be extended beyond θ, supposing that the probabilities defined before θ remain the same over time. Thus, for example, if an individual is not married in θ, we may suppose that his age at marriage is a random variable T following the same distribution function as that observed for the individuals already married who have similar characteristics. Besides, this hypothesis is sometimes unnecessary, and one can simply work on event histories that are completely terminated, as is done in historical demography.

The method of event history analysis we present here will thus involve estimating the probability distribution of the life-courses followed by a given population. This distribution may vary from one sub-population to another and may depend on certain characteristics of the individuals in the sub-population (social and economic characteristics of parents or grandparents, for example). These life-courses

---

[1] The set of the Ω_θ parts.

feature random variables $T_1, T_2, \ldots, T_n$ which represent the duration of stay in the different states that constitute them. Of course, these variables are not independent, and the distribution of life-courses to be estimated is the result of their joint distribution.

Our approach, therefore, supposes that individual behaviour can be described as a complex stochastic process. Having once assumed this model, we shall begin with a statistical estimation of the distribution of the variables previously defined which we develop in the present work. Once known, these distributions make it possible to deal with the more complex distribution of the overall life-course.

We shall first examine methods for collecting these event histories. Generally, it is not possible to have an exhaustive collection of biographies. More often, one works with data on partially observed event histories collected from a sample of individuals. It is therefore necessary to examine the different problems raised by such incomplete observation.

We shall then proceed to formalize the methods of analysis and estimation, starting with the most simple case and gradually introducing an increasing degree of complexity. After studying an event, we shall move on to the reciprocal study of the interactions between two events, before extending these non-parametric models to cover more complex situations. Next, we shall examine parametric models which make it possible to introduce the effect of a large number of characteristics on the duration of stay in a given state. Finally, we shall deal with semi-parametric models which combine the two preceding approaches.

These various methods will be illustrated by applications to very different situations, so as to show the possibility for their generalization. In the Appendix, we shall indicate the programmes for carrying out these analyses, so that the reader may actually make use of them.

This book provides both a detailed theoretical presentation of methods for event history analysis and a practical application of these methods to files that exist already or that are to be created on the basis of event history surveys.

I

Extending the Scope of Longitudinal Analysis

1
Observation of Event Histories

In the Introduction, we indicated the theoretical objectives of event history analysis. To achieve these objectives, it is necessary to have observations of the event histories in sufficient number to guarantee reliable results.

In this case, the ideal sample is an exhaustive observation of the population under study with the aim of providing a detailed event history of each of its members.

To this end, an exhaustive census can only put a limited number of questions about the event history. For example, a question about the place of residence on the first of January of the year of the previous census has been asked in France since 1962. This makes it possible to estimate the numbers of intercensal migrants, but does not take account of the date of migration, the multiple migrations during the intercensal period, or the return, resulting in an underestimation of mobility. Moreover, the data do not allow us to relate migrations to occupational or family events, for they are either incomplete or censored. Thus, for example, we know the marriage status of each individual, but we do not know the date of marriage (if married), of divorce (if divorced), etc. These data, therefore, cannot be used in event history analysis, for they concern only a tiny part of the life history.

It is more worthwhile to consider registration statistics, which are also exhaustive. They represent the basic source for the classic demographic analysis of phenomena taken separately. Nevertheless, the tables published by statistical institutes (INSEE in France) allow no analysis of the interaction between different phenomena, except to eliminate the disturbing effect of mortality. In any case, they concern only three demographic phenomena: nuptiality, fertility, and mortality. Finally, although they register the corresponding events, they do not accurately indicate which populations are at risk. In fact, the data obtained from a previous census should be used as the basis to estimate this population. Not taking the migrations into account biases the estimate, especially if only part of the country is considered. Again, these data prove to be insufficient for the purposes of event history analysis.

Panel data come from a permanent or almost permanent sample of individuals who are interviewed at regular intervals. The data thus collected, however, pose problems of analysis. In fact, since the individuals are asked about their situation only at the time of each interview, the individual event histories are incomplete, and these 'gaps' make it difficult to carry out a statistical analysis of interactions, without constraining hypotheses (cf. Section 3.3). Although the panel method is an alternative to a single-round survey, the temporal dimension of the information obtained is not always easy to use at the individual level. Yet these data, collected successively on the same sample, represent an immense wealth of information with which to measure the evolutions at the overall level. The INSEE Labour Force Survey is a panel survey made each year using a sample that is partially renewed each time.

The data of population registers, which are as exhaustive as registration statistics, are far more satisfactory when the registers are correctly kept. Apart from births, marriages, and deaths, they register the internal migrations which modify the population of an administrative unit. They do not take into account changes in occupational status or various events non-officially recorded, such as temporary consensual unions. It must also be noted that these registers exist only in a limited number of countries (there are none in France) and that their role is essentially administrative. When they are not centralized, it is very difficult to follow the progress of an individual through all the places where he has lived. It can be done only for small sub-populations. There is an example of such a follow-up in Sweden, where the individuals born in the Arnas parish in 1896–1905 were followed during 50 years of their existence (Wendel, 1953). In Belgium two such follow-ups were conducted, the first one on 50 couples born in the 1910s, throughout 60 years of their life (Duchêne, 1985), and the second one on 445 couples born between 1933 and 1942 (Poulain *et al.*, 1991).

Hence we see that, among the sources mentioned so far, the population register is by far the most suited to the purposes of event history analysis, yet it cannot be used exhaustively because of its prohibitive cost. On the other hand, it is completely feasible to follow a sample with the help of the register, leading to data of the same type as the data obtained from a survey, and probably of better quality, although less comprehensive. (Occupational data in particular is lacking.)

1.1. Various Types of Surveys

In a country where population registers are unavailable, surveys are necessary to carry out event history analysis. They are also useful in countries where population registers are available, to capture the non-recorded events. Several types of surveys can be undertaken.

A *prospective survey*, following individuals since their date of entry into the population at risk (14 years, for instance, if marriage is being studied; age at leaving school if professional careers are studied; etc.) presents the most favourable situation. A *multi-round survey* is of a prospective type, but it does not necessarily follow the individuals ever since their date of entry into the population at risk, and it loses them if they move out of the sample's natural area. In that case it is necessary to carry out a retrospective survey during the first round to be able to observe all the past existence of the respondents and to follow the individuals who move out of the sample's natural area. Under these conditions, the surveys will provide information on all the events experienced by the individuals up to the date of the last interview. But, as the individuals are not usually followed until their death, some intervals observed will be *right-censored intervals*. Thus, for instance, some women will have had their first child during the survey and we do not know whether they will have had a second one afterwards and, if so, the date of this second birth. We shall see later that the information provided by such a survey can nevertheless be used.

In a prospective survey, we must be careful about the loss of individuals in the course of the survey (individuals whose address cannot be found after a certain date). In this case, it is possible that the censoring time might not be random and might concern individuals whose behaviour pattern differs widely from that of the rest of the population. This hypothesis can be borne out by dealing separately with the sub-population that has been lost before the end of the survey, to check whether it does not already have a different behaviour pattern while still under study.

Thus, during a longitudinal three-round survey (1974, 1976, and 1979), A. Monnier obtained rather good results. The survey was aimed at comparing fertility plans with actual behaviour. For the survey, he re-interviewed the women of his initial sample primarily by post and succeeded in minimizing the bias due to losses: 82 per cent of the respondents answered in the second round and 89 per cent in the third (Monnier, 1987).

In a prospective survey, therefore, it is important to equip oneself with all possible means to trace a respondent who has out-migrated. Noting the names and addresses of people who know the person well has proved very useful in this respect.

A *retrospective survey* presents no such drawback. In fact, the individuals are interviewed only once, and are asked to give all the dates of occurrence of the events studied since their entry into the population at risk. Again, we shall have right-censored intervals, by the date of the interview. But as the date of the interview is independent of the dates of occurrence of the various events, no problem should arise when using the collected information.

Such a survey, however, involves some risks of bias which will be developed later. For example, there are the individuals who have died or out-migrated before the survey. For them, death or out-migration can very well be related to the event under study. Thus, for instance, if nuptiality is being studied, it is obvious that some illnesses or disabilities decrease the chances of a person's getting married and, at the same time, increase the risks of dying.

We must also take note of the memory problems of respondents. In a retrospective survey, we are led to interview individuals about events that occurred over 50 or even 70 years before. It may be assumed that the reliability of memory decreases with the length of time elapsed since the event recalled and that some dates are given with a major element of doubt or are even left out. It is important, therefore, to ascertain the quality of the data collected in a retrospective survey. In particular, in countries where population registers are available, it is possible to compare data obtained from retrospective surveys with data obtained from registers. Later we shall briefly present some results already obtained in Belgium (Poulain *et al.*, 1991) and others from Sweden (Lyberg, 1983). Such an analysis must be carried out on large numbers to guarantee the validity of the results obtained in a retrospective survey. The perception of time can vary to a great extent according to the society under study, and there can be different solutions to the same problem.

Apart from these two main types of survey, there are other types which lead to more complex situations and to far more important biases which we shall examine later.

Before considering them in detail, however, we shall briefly present the survey on the family, work, and migration history, 'Triple Biographie', which we carried out at INED in 1981 and which will provide many examples of application of the various methods of analysis presented in this book.

1.2. The 'Triple Biographie' Survey

This survey belongs to a tradition of experiments which has existed for a long time at INED.

The first survey of this type was carried out in 1961 by Guy Pourcher on the Paris population (Pourcher, 1964). He had interviewed a sample of Parisians on various stages of their past life, providing a first model of questionnaire. The classic event history analysis he made of it, however, left unused a great part of the collected data (dates of the various events, in particular).

The 'Triple Biographie' survey was carried out to make an analysis of interactions between various aspects of the family, work, and migratory life of the respondents. This analysis affected the choice of the sampling plan and of the questions asked.

This is a statistical, biographical, and retrospective survey of national scope, concerning all individuals born between 1911 and 1935, undertaken in 1981. These individuals were included without any selection based on their life history. It would have been better to carry out the survey on several generations, for instance those born in 1911, 1921, and 1931. Each of the resulting groups of individuals would have experienced the various economic events (crisis, wars, etc.) at exactly the same age. Unfortunately, this restriction is not easy to make in a country where no population register is available and where access to sampling databases is particularly restricted. Thus, the 1975 Law 'Informatique et Libertés' prohibits for 100 years the use of a file for other purposes than its original use. Obviously, in countries where population registers are available and can be used to create a sampling database, we strongly recommend observing precise generations. This was done, for instance, in the survey carried out by the Max-Planck Institut of Berlin under the supervision of Karl Ulrich Mayer: three cohorts born in 1929–31, 1939–41, and 1949–51 were observed.

To carry out this survey in France, we used as our sampling database the register of the dwellings existing in 1975 or completed since, in the *communes* (French administrative districts) of the INSEE 'master sample'. The fact that only 45 per cent of the households included an adult aged between 45 and 69 would have implied too many useless visits, very discouraging for the interviewers. An original solution to this problem consisted in using the same sample of dwellings for another survey, on the 'family and work history' of women with one or more dependent children. The two sub-populations concerned are almost separate: only 10 per cent of households include both a child under 16 and an adult aged between 45 and 69. Actually, if the two samples had been entirely disconnected, the sampling scheme would have been truly non-informative on the life histories collected. This drawback, however, is of minor importance here. On the other hand, that solution made it possible to limit useless visits and made the survey much more worthwhile, which is why we finally chose it.

The data for the two surveys were therefore collected simultaneously from the same sample of 16,500 dwellings. The households concerned by the two surveys were allocated at random to one or the other of the surveys by using the parity of the number on the address sheet. When the household included several persons aged between 45 and 69, the person interviewed was designated using the Kish method.[1] These surveys were carried out with the help of the INSEE regional offices and their interviewers: only 11 per cent of the selected dwellings resulted in a failure (refusal, long-term absence, occupant impossible to reach). In addition, the surveys left out 17 per cent of empty dwellings or second homes and 20 per cent of households out of the survey coverage (Riandey, 1985). Thus, 4602 questionnaires were completed with 'family, work, and migratory histories'. The interviews lasted one hour and ten minutes on average, with some lasting over two hours.

We shall now examine the content of the questionnaire.

To analyse the interactions between the various events, the precise date and the localization of the events were collected. Thus, concerning the individual's first marriage, the following questions were put to him:

- When did you marry (month, year)?
- What was your place of residence after your marriage (town, *département*, or country)?

---

[1] The Kish method makes it possible to select one person among several without leaving any personal initiative to the interviewer, thanks to a table where the two entries are the number of eligibles and a 'Kish' number is attributed at random to the household.

Concerning a period of employment, the following questions were asked (among others):

- At what date did the period of employment begin (month, year)?
- Where is the firm located (town, *département*)?
- What is the precise activity of the firm?
- What was your (very precise) main occupation at the beginning of the period?
- What was your main occupation at the end of the period?
- If the period is ended, at what date (month, year) did it finish?

Finally, for successive dwellings, the following information (among others) was requested:[2]

- When did you move in (month, year)?
- Where is the dwelling located (*commune*, *département*)?

Even though the dates were given to within a month, this information did not prove to be very reliable, for many questionnaires did not include it. This appears clearly when compared with data obtained from population registers (Poulain *et al.*, 1991).

To define a status accurately, it is also necessary to collect as much information as possible on its characteristics. Thus, for employment 12 characteristics are collected, which make it possible to reconstruct the INSEE code of occupational status or to define precise categories (whether employed in the public or the private sector, for instance). Similarly, for each dwelling information is gathered on the residential status in respect of five items, which enables us to classify the individuals in different ways (owner or non-owner, for example).

In such a questionnaire, we must not forget to include information about the family origins and the brothers, sisters, and children of the respondent, which completes the purely biographical part.

It is also essential to be able to compare the event history of the respondent with that of his/her spouse. Initially, we wished to interview both spouses in order to collect further information on the family origins, the work history, or the pre-nuptial migrations of the spouse. But the second interview, though shortened to a half-hour, met with many refusals during the pilot survey. As the subject of the second interview was often the spouse of a housewife interviewed first, it

---

[2] The reader who is interested by the questions asked on the periods of employment, inactivity, and successive dwellings will find this part of the questionnaire reproduced in the textbook *Méthodes de mesure de la mobilité spatiale* (Courgeau, 1988).

could have led to important biases. To avoid the bias, we abandoned the two interviews in favour of a Kish sample. However, the respondent was asked to provide a great deal of information on the origins and the work history (at the time of marriage and of the survey) of his/her spouse.

In this form, the questionnaire has already made it possible to carry out many analyses of the interactions between phenomena (Courgeau, 1984*a*, 1985*a*, *b*, 1987*c*; Courgeau and Lelièvre, 1986, 1988*b*; Lelièvre, 1987*a*, *b*).

1.3. The Problem of Weight

We have seen that only one of the household members aged between 45 and 69 was interviewed in the 'Triple Biographie' survey. It seems *a priori* useful, therefore, to weight the sample in relation to the number of individuals represented by each respondent. We shall show that, in order to carry out an analysis according to the models presented in this work, such a weighting does not apply, for the sampling plan does not provide any information on the collected life histories.

First, we shall give a concrete example of the results obtained by using or not using weighted data.

Let us evaluate the impact of age at the beginning of the observation period and the impact of the duration of stay on the probability, for active men from the generation born between 1926 and 1935, of experiencing a change of dwelling. To do that, we assume that a multiplicative Gompertz model applies to the data—a model developed in Section 7.1.3 below. Here, we shall only mention that, if the model is verified, the migration rate is written

$$h(t; z) = \exp(z\alpha + \beta t),$$

where z is a row vector of individual characteristics (here, the unit for the constant and binary variables equal to the unit if the individual begins to be observed in the age group concerned), α is a column vector of parameters which indicate the impact of these characteristics on the rate, t is the duration of stay, and β is a parameter indicating the impact of this duration.

Table 1.1 reports the estimates of these parameters, whether the data are weighted or not. The weighting is important, for it almost

TABLE 1.1. The effect of age and survival time on the change of dwelling of active men of generations born between 1926 and 1935

Individual characteristics	Model without weighting		Model with weighting
	Estimated effect	Standard deviation	Estimated effect
Constant	−2.258	0.0607	−2.263
Aged under 20	0.612	0.0671	0.627
Aged 20–24	0.553	0.0686	0.573
Aged 25–29	0.356	0.0740	0.309
Aged 30–34	0.250	0.0827	0.216
Duration of stay (β)	−0.056	0.0033	−0.054
No. of individuals	3429		6315

doubles the number of individuals to be taken into account. On the other hand, it has a negligible impact on the estimated parameters, when the standard deviation of these parameters is taken into account in the model without weighting. Thus, for instance, an individual aged under 20 will have a migration rate of 0.110 after a ten-year observation period in the model without weighting, and of 0.113 in the model with weighting.

We shall not reproduce here the detailed demonstration given by Hoem (1985) showing that weighting is not to be introduced under much more general conditions. However, we shall indicate the precise conditions which lead to ignoring the sampling plan when analysing event histories.

For this result to be verified, the main condition is that the sampling plan should be independent of the life histories of the respondents. Also, the sampling plan must be non-informative. Thus, in the survey made in France individuals interviewed were selected by the Kish method. Yet this method does not take account of any information on the event history of the persons to select, since the 'Kish number' was attributed at random to the household. Consequently, the sampling plan is truly non-informative.

Another condition is that the observation of individual event histories, which can be partial, should be made independently for each individual. This condition is met, in particular, if the life history is observed for a prespecified duration of marriage or unemployment which is identical for all the respondents.

If these two conditions are met, then it can be shown that the inclusion of weights is not needed in event history analysis. Event histories would have the same probability distribution if the total population were observed in an exhaustive census. In that case, it is also said that the sampling plan should be ignored. This result does not depend on which model is used and remains valid even if it is misspecified.

The conditions leading to this result can cease to be met if data are collected from a retrospective survey. In such a survey, only those individuals who are still living when it is carried out are interviewed. Those left out by the survey may have had a demographic behaviour very different from that of the survivors. If this is true, the individuals observed have been subjected to a selection process which makes the sampling plan informative. On the other hand, if the persons who have died or emigrated abroad had the same behaviour as the survivors during their life in the country, then the selection bias entirely disappears: the sampling plan becomes non-informative (Hoem, 1985).

We know that in France, as in other countries, the mortality of single persons is much higher than that of married individuals of the same age. Thus, for instance, in 1967–9 in France the mortality rate of bachelors aged 45–49 was 612 per 10,000, whereas the mortality rate for the married men in that age group was only 298 per 10,000 (Vallin and Nizard, 1977)—that is, less than half the rate for bachelors. Such a selection introduces errors in the rates that are estimated through a retrospective survey, for which the sampling plan becomes informative. Nuptiality rates will be overestimated by the survey.

In that case, it is possible to correct these errors if vital statistics or, better still, population registers are available. We can then introduce a weighting different from that due to the sampling plan. In order to do that, it is necessary to know the probabilities of inclusion of the various kinds of event histories in the sample.

1.4. Incomplete and Erroneous Survey Observation

We have previously indicated some imperfections in surveys. Let us examine these sometimes unsolvable problems in detail.

Here it is useful to give some examples of complex situations arising from the method of collect. Suppose that the individual is interviewed about the various changes of residence he has experienced or the various births that have occurred during the last five years. It appears that, if in some cases there always are right-censored intervals, there are also left-censored intervals and even some intervals where both the beginning and the end are unknown. Such data cannot be used unless very strong hypotheses are introduced about the distribution of events over time. Thus, assuming that events follow an exponential distribution, the left-censored intervals do not produce any bias. In effect, we shall see (Section 7.1.1) that the conditional distribution of $(T - t_0)$, knowing that T is greater than t_0 (this date being situated five years before the survey, for example), is the same as the distribution of T. However, it is very unlikely that this condition should be verified. Consequently, we advise against using this method of collect when no other information is available on the phenomenon under study.

It is possible to avoid such a bias by asking a question about the date of arrival in the dwelling occupied five years before. This was the solution chosen in the survey on residential mobility between 1973 and 1978, a side-survey to the INSEE Labour Force Survey. But in this case the mobility previous to that date, which could affect the observed mobility, is still not taken into account. Similarly, surveys that interview on the previous change of status (for instance, last migration, last occupational change, last birth,) lead to important risks of bias.[3] Thus, this observation of durations of stay includes more longer intervals, which are known to be subject to length bias. When other data on duration of stay are unavailable, it is not at all advisable to use this method of collect. In fact, by asking about an individual's last migration, we know that this event will not occur again before the survey. Therefore, the sampling plan is informative. More generally, we can say that these selection problems are due to the individual's behaviour pattern being observed under conditions that affect it. In all these cases, the observation is not neutral towards the behaviour pattern under study, and that could introduce biases. To analyse these data, it is necessary to formulate hypotheses and to test their validity.

In a prospective survey, it is often difficult to achieve the follow-up of the individuals. This is why, too, a selection might be at work: the non-responses and the losses in the course of the follow-up can occur preferentially in a specific category of the interviewed population.

---

[3] For further details on this subject, see Hoem (1985).

TABLE 1.2. Fertility rates of parity 1, based on data for the total population (h_P), the entire target sample (h_S), and the respondents (h_R) per 1000 women per year

Age	h_P	h_S	h_R	Estimated non-response bias
23	138	$120 \pm 27^*$	$128 \pm 30^*$	8
24	152	192 ± 36	219 ± 42	27
25	165	158 ± 36	163 ± 40	5
26	168	142 ± 37	154 ± 42	12
27	158	145 ± 40	151 ± 45	6
28	142	129 ± 40	133 ± 45	4
29	125	155 ± 47	177 ± 57	22
30	107	107 ± 42	133 ± 53	26

* Confidence interval at the 95% level.

Source: Lyberg (1983): 1981 Swedish fertility survey, women born between 1941 and 1945.

There can also be a conditioning bias (Deroo and Dussaix, 1980) arising from a selective change in the behaviour pattern of the individuals concerned. Whether or not this bias exists, however, is the subject of a controversy. Such a follow-up is now being carried out by F. Cribier, who is observing the whole of a cohort of Parisians from their retirement in 1972 to their death.

Refusals to answer to prospective as well as retrospective surveys are another source of error. A study made with the Swedish 1981 fertility survey enables us, through a comparison with the data obtained from population registers, to assess the importance of such errors (Lyberg, 1983). The rate of non-response to the survey was 13 per cent, and the individuals who refused to answer it could have come from a very selective group with an atypical behaviour pattern. The results of the comparison show that this was not so—various analyses made from the register and retrospective survey data give very similar results (Table 1.2).

Memory errors can be important, especially when interviewing elderly people through retrospective surveys. The dates of various events can be wrong, or even forgotten. It is necessary, therefore, that the quality of these data should be tested by comparing them with those obtained from population registers. Such a test has been carried out in Belgium on the 'Triple Biographie' questionnaire, in a country where the data from population registers make such verification possible (Poulain et al., 1991). We shall now present these results in more detail.

The test was carried out on a sample of 445 couples,[4] by putting the respondents in the most unfavourable conditions. At first, both spouses were interviewed simultaneously in separate rooms. After the interview, the two spouses confronted their respective accounts of the events in their life and corrected it, in particular referring to any document that was available (family record book, rent receipts, etc.). Lastly, the registers were consulted, independently of the interview, as a fourth source of information on the same events. The sample was limited to couples in which one spouse was born between 1933 and 1942 and the other in a larger time-frame, between 1933 and 1947. The range of ages at the time of the survey, therefore, was 41–55. These respondents were asked to recall events that sometimes had occurred very long ago. The comparison of the dates given by each of the spouses and by the couple together, along with those provided by the population register, is reported in Table 1.3 (see Poulain et al. 1991 for further details). This is satisfactory for the date of marriage and the birth or death of a child, for these are registration records, whereas it is less so for the migrations of the household. In fact, though it is a legal obligation to declare the change of dwelling within an eight-day period, there may be some delay—which only rarely, however, exceeds a month. Moreover, population registers do not record the changes of dwelling abroad, and record only partially the out-migrations (in-migrations) towards (from) a foreign country, particularly in the case of military men or people serving on Voluntary Service Overseas.

Table 1.3 reveals a very distinct difference between marriage and the birth of the children on the one hand (dates are exact to within a month in over 90 per cent of cases) and the migration of the household or the emancipation of the children on the other (dates are exact to within a month in only 39–67 per cent of cases). However, these last percentages increase if we choose an interval of more or less than one year: regarding the emancipation of the children, it rises to 69.8 per cent for men, to 76.7 per cent for women, and to 78.4 per cent when the spouses are together; regarding migrations, it rises to 87.5 per cent for women, to 90.2 per cent for men, and to 93.2 per cent for the two

[4] A test limited to 50 couples had already been made before (Duchène, 1985; Courgeau, 1985a), leading to results very similar to those presented here.

TABLE 1.3. Omissions and errors in dates of marriage, birth of the children, emancipation of the children, and household migrations*

	Marriage			Birth of children			Emancipation of children			Household migration		
	Men	Women	Couple	Men	Women	Couple	Men	Women	Couple	Men	Women	Couple
Total no. of events	445	445	445	1078	1078	1078	310	310	310	1388	1388	1388
No. of dated events	440	445	445	1076	1077	1078	222	228	222	1169	1196	1237
of which												
Exact date (to within 1 mo.)	414 (93%)	440 (98.9%)	443 (99.6%)	940 (90.8%)	1038 (97.8%)	1045 (98.2%)	83 (39.0%)	115 (50.4%)	119 (53.6%)	651 (55.7%)	723 (60.5%)	833 (67.3%)
Antedated by 1 yr or less	8 (1.8%)	1 (0.2%)	0 (0.0%)	46 (4.4%)	6 (0.5%)	8 (0.7%)	42 (20.0%)	39 (17.1%)	38 (17.1%)	269 (23.0%)	273 (22.8%)	257 (20.8%)
Postdated by 1 yr or less	13 (2.9%)	3 (0.6%)	1 (0.2%)	33 (3.2%)	10 (1.0%)	8 (0.7%)	23 (10.8%)	21 (9.2%)	17 (7.7%)	103 (8.8%)	83 (6.9%)	63 (5.1%)
Antedated by over 1 yr	2 (0.4%)	0 (0.0%)	0 (0.0%)	7 (0.7%)	3 (0.3%)	3 (0.3%)	40 (18.9%)	36 (15.8%)	34 (15.3%)	100 (8.6%)	83 (6.9%)	53 (4.3%)
Postdated by over 1 yr	3 (0.7%)	1 (0.2%)	1 (0.2%)	9 (0.9%)	4 (0.4%)	1 (0.1%)	24 (11.3%)	17 (7.5%)	14 (6.3%)	46 (3.9%)	34 (2.9%)	31 (2.5%)

* Proportions in relation to events concerned appear in parentheses.

Source: Poulain et al.

spouses. In fact, the information provided by the women is always better than that given by the men, and the information provided by the two spouses is always better than that given by either the women or the men. Consequently, it is always preferable, where possible, to interview the two spouses together, or, as a second-best solution, the women, on both family events and migrations.

Yet these errors of date hardly affect the results of the analyses, whether parametric, non-parametric, or semi-parametric, which we have carried out using these different data (Courgeau, 1991). Here we present the results of a parametric analysis of the durations of stay of over six months in the different dwellings.

Table 1.4 reports the results of this analysis, in which we have taken account of the duration of marriage and the time spent in the dwelling, as well as the number of children born at the beginning of this period. In a second model we also took into account the residential status indicated by the men, the women, and the two spouses together (this information is not recorded in population registers).

The model chosen implies that the impact of the duration of stay decreases exponentially the migration rate, which is written as before at time t:

$$h(t; z) = \exp(z\alpha + \beta t),$$

where z is the vector of the covariates presented before, and α and β are the parameters to be estimated which will give the impact of the characteristics and of the duration of stay.[5] Table 1.4 reports the results of the first model (excluding the characteristics of the residential status), estimated separately using the data given by the men, the women, the two spouses together, and the population registers [6]. All the characteristics have a very similar effect, whatever the source of information. The number of children at the beginning of the period has no effect on the duration of stay in the dwelling, whereas the duration of stay itself and the duration of marriage at the beginning of the stay strongly influence the migration rate.

The table also reports the results of the second model, in which, apart from the previous characteristics, the residential status was also taken account of. At first, we observe a very significant effect of the residential status whatever the source used, all the estimates being in

---

[5] See Sect. 7.1.1 and 7.3.1 for a more detailed presentation of this type of model and of the methods of estimation of the parameters.

[6] The parameters were estimated using the RATE program developed by N. Tuma.

TABLE 1.4. Analysis of spatial mobility: effect of the time elapsed since marriage, of the duration of stay in years, and of the residential status on the probability to change dwelling according to the source[a]

Variables	Men (1260 stays)		Women (1310 stays)		Spouses together (1314 stays)		Registers (1189 stays)
	Model 1	Model 2	Model 1	Model 2	Model 1	Model 2	Model 1
Constant	−2.955*** (0.174)	−2.322*** (0.193)	−3.009*** (0.179)	−2.391*** (0.193)	−3.062*** (0.183)	−2.264*** (0.192)	−3.118*** (0.187)
Beginning of the stay on year of marriage	1.464*** (0.172)	0.438*** (0.174)	1.569*** (0.177)	0.574*** (0.178)	1.642*** (0.182)	0.575*** (0.181)	1.564*** (0.186)
Beginning of the stay between 1 and 4 yrs. after marriage	1.117*** (0.160)	0.572*** (0.160)	1.198*** (0.157)	0.622*** (0.158)	1.301*** (0.161)	0.750*** (0.161)	1.199*** (0.166)
Beginning of the stay between 5 and 9 yrs. after marriage	0.641*** (0.164)	0.375** (0.165)	0.559*** (0.162)	0.286* (0.163)	0.684*** (0.164)	0.444*** (0.165)	0.707*** (0.166)

TABLE 1.4. (continued)

No. of children at the beginning of stay	0.006 (0.040)	−0.016 (0.039)	0.052 (0.042)	0.042 (0.041)	0.051 (0.042)	0.023 (0.042)	0.033 (0.044)
Housed by the employer		0.485*** (0.091)		0.480*** (0.090)		0.321*** (0.078)	
Home-owner		−2.431*** (0.175)		−2.347*** (0.166)		−2.538 (0.164)	
Duration of stay	−0.113*** (0.0077)	−0.056*** (0.0076)	−0.116*** (0.0076)	−0.058*** (0.0075)	−0.119*** (0.0077)	−0.064*** (0.0075)	−0.104** (0.0076)

[a] Parametrers estimated; standard deviations given in parentheses.
* significant result at 10% level.
** significant result at 5% level.
*** significant result at 1% level.

the same confidence intervals. The impact of the other characteristics is limited, but remains similar to what it is when they are taken into account alone. This analysis reveals that important errors on the dates of migration and the durations of stay do not produce significant biases in the analysis of migration rates according to the various characteristics of the respondent at the beginning of the stay. In most cases the results are consistent whatever the source of information used: the few differences that are observed do not modify the main results of the analysis.

Other analyses have been carried out, in particular on the relationship between the birth of the first child and the first migration after marriage (Courgeau, 1991), giving results that are still very consistent. Nevertheless, the poor quality of the answers provided by the men on the age at marriage of their wives diminishes the effect of that age when it is introduced as a characteristic in the analysis.

As a conclusion, we may say that, although the errors concerning the dates of the various events can be important, they hardly seem to modify the logical order in which these events occur. This order is correctly recalled, and the memory errors provide a background noise from which coherent information can be extracted whatever the source used. Memory seems reliable, therefore, where the analysis requires it. However, it is always necessary to collect the information directly from the person who has experienced the event, rather than from a third party. It is also preferable to collect the information from the women, or better still from the two spouses together, when making a retrospective survey, and to induce them to confront the dates given with all the documents available in the family.

1.5. The Problem of Censoring

One of the major difficulties in event history analysis originates in the presence of censoring, which indicates gaps in information that exist each time part of the event history is missing.

It is said that there is a *left-censoring* when the history starts at one moment in the life cycle and no information about the previous period is available. For example, collecting the migrations of the residents in a geographic area from a given date does not take account of their previous moves, in particular the moves that led them to the area considered, unless a preliminary retrospective survey completes the collection. Similarly, the study of individuals' work histories within a company leaves out their previous work histories. We shall then consider implicitly that the previous work history, outside the present company, has no impact or a negligible impact on the career within the group under study, and shall attach more importance to the training of individuals, for instance, as an explanatory factor.

Conversely, there is *right-censoring* when the account available for each individual is interrupted, either because the individual has disappeared from the sample, or because the account stops at the time of the interview. Whatever the method of collect, whether prospective or retrospective, the biographical data characteristically involve censoring at the end of the observation period, unless the prospective is prolonged until the death of the respondents. In a purely prospective survey, there can be both right- and left-censoring. In a retrospective survey, the date of the interview is also the date at which the account ends, beyond which no information is collected.

As this censoring is unavoidable in biographical data, one of the aims in analysing the durations of stay is to take systematically censored data into account and to control them (Kaplan and Meier, 1958).

The demographer is confronted with a file recording, for each individual, temporal series concerning family life, work history, migrations, etc. This therefore provides the demographer with chronological sequences punctuated by various kinds of occurrences, in different orders, according to the individual and not necessarily present in every biography.

If we consider three different events—departure from the parents' residence, marriage, and the first job—their order of occurrence in the life of an individual can vary in six different ways. The number of possibilities is considerably higher if we take account of the cases when one of the events has never occurred (single people, housewives, farmers still living with their parents); when two events never occur (single persons living with their parents, single persons without occupation, married people without occupation and living with their parents), and finally where none of the three events is experienced.

These complex life-courses constitute the material for the analysis. More than the age of the individuals, these temporal milestones represent the transition from one state to another. They mark the adjustments, the choices, whether objective or not, that individuals make in the course of their lives. Thus, these occurrences are evidence of the 'social age' of individuals, the concept of which has been formalized by the sociological literature (Elder, 1978; Foner and Kertzer, 1978; Hogan, 1978), which has also described its evolution (Modell *et al.*, 1978) and sexual nature (Langevin, 1986). It is therefore necessary to modelize the occurrence of these events by considering only one of them, or else one that is repeated (successive births), or several events in interaction.

1.6. Conclusion

As we saw, collecting these event histories from population registers as well as biographical surveys raises a certain number of problems.

The most satisfactory source is undoubtedly population registers, at least in countries where they are adequately kept. Unfortunately, there are only a few such countries, and this source of data is very expensive to use. Moreover, it does not record some events or some pieces of information that are necessary for event history analysis. Thus, to carry out a detailed analysis of Swedish women's fertility, it has been necessary to undertake a retrospective survey, though there are perfectly kept registers available in Sweden.

Prospective surveys also represent a very reliable source, when successive surveys are carried out at not-too-long intervals (one year, for instance): otherwise, the time necessary to obtain data that can be analysed is very long and this limits the use of the source, which must often be associated with a preliminary retrospective survey.

Retrospective surveys, therefore, remain the most used source in event history analysis. But they raise many problems that are not yet totally solved. Thus, the selection of individuals surviving at the time of the survey can make the sampling plan informative. It is therefore important to assess the biases introduced and to check that they are not significant. Similarly, these surveys will give rise to problems of memory, especially when interviewing elderly people. Errors can be very important, but it may be assumed that the biases they introduce in the analysis are much more limited. Comparing the results obtained through retrospective surveys and population registers should make it possible to answer these questions.

In spite of these drawbacks, however, retrospective surveys represent the easiest method of data collection. It is important, therefore, that the results they provide us with should be well ascertained wherever possible.

2

Formalization of the Analysis

We shall now deal with the formalization of event history analysis.

This formalization extends the methods developed by biomathematicians for the analysis of mortality, in the presence of restricted samples, to the 'analysis of the durations of stay', which does not assume that the studied event is death. Contrary to death, it is entirely possible that marriage, the birth of a child, a migration, or an occupational change may never occur during the life-course of a given individual.

The observation of event histories provides us with a certain number of occurrences, the distribution of which we want to study over time. In order that these occurrences can be precisely defined, three conditions must be met.

1. A common and clearly defined origin must be available. This moment is not necessarily marked by the birth date of each individual: more often it corresponds to the date of occurrence of a previous event (date of first job, of marriage, etc.), after which the events under study may occur (change of jobs, birth of legitimate children, etc.).

2. The time must be marked on a common time-scale: time elapsed since the outset, age of the individuals.

3. The occurrence itself must be clearly defined, which is generally the case for demographic events when their dates are being collected. However, it is necessary to specify clearly the type of events taken into account. Thus, migrations can be defined in very different ways, which should be well specified (all the moves or only intercommunal migrations, for instance, which eliminate intracommunal migrations). The same can be said for occupational changes (detailed occupational changes, changes in occupational qualification, changes in social status category).

For greater clarity, we shall start with the simplest case—studying a population in which all individuals have the same probability of experiencing a given occurrence at any moment. This probability, however, may change over time.

From this simple theoretical case, we can introduce a more complex heterogeneity of the population under study and take into account a growing number of types of occurrences to be analysed simultaneously. We shall then show how the models presented here represent a generalization of differential demography and Markov models.

2.1. Analysis of a Homogeneous Cohort Experiencing a Single Event

This simple case corresponds to the demographic analysis of a single event. Compared with classic demography, it introduces no new element apart from a specification which will be applicable to more complex cases.

2.1.1. Failure time distribution, notation, formalization

Let us consider a homogeneous population whose individuals are likely to experience a given event at time T. T is therefore a positive random variable, the distribution of which we shall now examine.

Time can be specified in various ways, but in the context of our analysis three kinds of specifications prove to be the most useful: the survivor function, the density probability function, and the hazard rate, defined as the density probability function conditional on the stay in the initial state.

The survivor function is defined, whether the distribution is discrete or continuous, as the probability of T being at least as great as a given value t:

$$S(t) = P(T \geq t), \quad 0 < t < \infty. \tag{1}$$

This function is clearly monotone non-increasing left continuous with $S(0) = 1$ (and only if all the individuals in the sample under study experience the event: $\lim_{t \to \infty} S(t) = 0$).

In practice, not all the individuals experience the event, but this characteristic of the survivor function limit is often necessary. To introduce those individuals who never experience the event, we assume the existence of a limit point ($t \to \infty$) such as $\lim_{t \to \infty} S(t) = 0$.

Generally, the distribution of T can have continuous or discrete components.[1] We shall first indicate its terms in both cases.

---

[1] This is discussed in the following section.

Formalization of the Analysis

If the distribution of T is continuous, the density probability function is defined as the limit when $\Delta t \to 0$ of the probability of a failure T being contained in the interval $[t, t + \Delta t]$ divided by Δt:

$$f(t) = \lim_{\Delta t \to 0} \frac{P(t \leq T < t + \Delta t)}{\Delta t}, \tag{2}$$

or otherwise

$$f(t) = -\frac{dS(t)}{dt} = -S'(t), \tag{3}$$

and therefore, conversely;

$$S(t) = \int_t^\infty f(s)ds, \tag{4}$$

and, whether all individuals experience the event or not,

$$\int_0^\infty f(s)ds = [-S(s)]_0^\infty = -0 + 1 = 1.$$

The hazard function specifies the instantaneous rate of failure at $T = t$ conditional upon survival to time t by

$$h(t) = \lim_{\Delta t \to 0} \frac{P(T < t + \Delta t \mid T \geq t)}{\Delta t}, \tag{5}$$

or otherwise

$$h(t) = \frac{f(t)}{S(t)} = \frac{-dS(t)/dt}{S(t)} = -\frac{d \log S(t)}{dt}. \tag{6}$$

$h(t)$ is truly a specification of the distribution of T. By integration, using $S(0) = 1$, we obtain

$$S(t) = \exp\left(-\int_0^t h(s)ds\right) = \exp(-H(t)) \tag{7}$$

and

$$f(t) = h(t) \exp\left(-\int_0^t h(s)ds\right) = h(t) \exp(-H(t)). \tag{8}$$

Note that $H(t)$ is the integral of the conditional density and is called the integrated or cumulative hazard function.

In the case where T is a discrete random variable and takes the values $t_1 < t_2 < \ldots$ with the probabilities $f(t_i) = P(T = t_i) = f_i$, then to stay in a given state until $T = t$, $(P(T \geq t))$, it is necessary and sufficient to move out of it at or just after t, so that

$$S(t) = \sum_{i \mid t_i \geq t} f(t_i) = \sum_{i \mid t_i \geq t} f_i. \tag{9}$$

The hazard function h_i is the conditional probability of the failure at t_i; that is,

$$h_i = P(T = t_i \mid T \geq t_i) = \frac{f_i}{S(t_i)}. \tag{10}$$

Lastly, a necessary and sufficient condition is to have never left at any of the dates (here discrete mass points) preceding t, so that

$$S(t) = \prod_{i \mid t_i < t} (1 - h_i). \tag{11}$$

In order to have $S(t) = \exp(-H(t))$ also in the discrete case, we take by convention

$$H(t) = \sum_{i \mid t_i < t} \log(1 - h_i) \tag{12}$$

as a cumulative hazard and, if the h_i are small;

$$H(t) \simeq \sum_{i \mid t_i < t} h_i. \tag{13}$$

2.1.2. The likelihood function

In each particular case, we intend to discuss the methods of calculation and of maximization of the total or partial likelihood. However, in this chapter on the formalization of the analysis, we present the likelihood function, to which the censoring mechanisms have given a particular shape. As before, we shall distinguish the case where the survivor function is continuous from that where it is discrete.

With a continuous survivor function, observing a failure at time t contributes to the likelihood (that is, the probability to observe the occurrences as they were effectively collected) by a term $f(t)$ that is

Formalization of the Analysis

a probability density of failure at t. On the other hand, if the observation is censored in τ, its contribution to the likelihood corresponds to its survival probability beyond τ, that is $S(\tau)$. The total likelihood for N individual and independent observations indexed by j is therefore

$$L = \prod_t f(t_j) \prod_\tau S(\tau_j). \tag{14}$$

In the case of a discrete survivor function, where at each mass point t_i corresponds to the probability $f_i = P(T = t_i)$, and where the contribution of the failure observation is f_i and that of a censored observation is, by convention,

$$P(T > \tau) = S(\tau^+) = 1 - \sum_{i \mid t_i \leq \tau} f_i, \tag{15}$$

as previously, we then have

$$S(\tau^+) = \prod_{i \mid t_i \leq \tau} (1 - h_i) \tag{16}$$

and

$$f_i = h_i \prod_{k < i} (1 - h_k). \tag{17}$$

It then follows that the total likelihood is obtained by calculating d_i, the number of occurrences at each mass point t_i and N_i, the number of individuals at risk at this same point. The contribution to the total likelihood is then

$$L_i = h_i^{d_i}(1 - h_i)^{N_i - d_i}, \tag{18}$$

so that the total likelihood is

$$L = \prod_i h_i^{d_i}(1 - h_i)^{N_i - d_i}. \tag{19}$$

This corresponds to the likelihood that would be obtained in the presence of a series of independent binomials with N_i trials and a probability of success h_i.

2.1.3. Discrete time, continuous time

By formalizing the above analysis in its more general terms, we were able to measure the proximity of the estimators obtained according to whether T is a discrete or a continuous random variable. This is more obvious if we specify the survivor function in the case of a mixed distribution. In fact, by designating h_c as the hazard function for the continuous part and $x_1 < x_2 < \ldots$ as the support points for the discrete part, the overall survivor function is

$$S(t) = \exp\left(-\int_0^t h_c(s)ds\right) \prod_{j \mid x_j < t} (1 - h_j). \tag{20}$$

It then follows that

$$h(t)dt = h_c(t)dt + \Sigma h_j \delta(t - x_j)\, dt, \tag{21}$$

where δ is the Dirac delta function:

$$\delta(x) = \begin{cases} 1 & \text{if } x = 0; \\ 0 & \text{if not.} \end{cases}$$

The cumulative intensity is then

$$H(t) = \int_0^t h(s)ds = \int_0^t h_c(s)ds + \sum_{j \mid x_j < t} h_j, \tag{22}$$

where the Dirac delta components define the discrete contributions to the integral. The survivor function may then be written in the discrete, continuous, or mixed case as

$$S(t) = \mathcal{P}_0^t (1 - dH(t)), \tag{23}$$

where the product integral $\mathcal{P}$ is defined in a way similar to a Riemann integral. The $[0, t]$ interval is divided into n small intervals $[0, t_1[$, $[t_1, t_2[\ldots [t_{n-1}, t[$. We then consider the limit when $n \to \infty$ and $\max(t_{j-1} - t_j) \to 0$ of

$$\mathcal{P}_0^t (1 - dH(t)) = \lim \prod_1^n (1 + H(t_{j-1}) - H(t_j)). \tag{24}$$

We can then write that

$$S(t) = \mathcal{P}_0^t (1 - h(u)du), \tag{25}$$

where again the Dirac delta functions handle the discrete contributions, so that, for the continuous part of the conditional density, we have

Formalization of the Analysis

$$\overset{t}{\underset{0}{\mathcal{P}}}(1 - h_c(s))ds = \exp\left(-\int_0^t h(s)ds\right). \tag{26}$$

It is therefore clear that in both cases—discrete time and continuous time—we are confronted with the same problem.

The methods in continuous time are those that suppose that the failure time is accurately measured. Actually, whatever the temporal division, the occurrences are measured in discrete units, and when those are small enough[2] the problem is treated in continuous time.

The rate in discrete time measures the probability that an individual experiences the event under study, knowing that he is at risk on this date. This rate is an unobserved variable, though it controls the occurrence and the chronology of events. We see here how important this variable is in a model of event history analysis.

In discrete time, we will divide the number of events at t by the number of individuals at risk just before t.

In continuous time, the above definition is no longer valid to characterize the hazard rate. In fact, in continuous time the probability that an event occurs exactly at time t is infinitesimal. Hence we must divide this probability during the $[t, t + \Delta t[$ interval knowing that the individual was at risk just before t, by the duration Δt.

Here it is clear that if $\Delta t = 1$ we again have the discrete hazard rate. And as was shown previously in (5) (the hazard rate), the continuous conditional density is given by

$$h(t) = \lim_{\Delta t \to 0} \frac{P(T < t + \Delta t \mid T \geq t)}{\Delta t}.$$

Although the hazard rate is usually referred to as the instantaneous probability of occurrence of the event under study, this rate has no upper limit. Interpreted more strictly, it also represents the unobserved rate at which the events occur, which, intuitively, illustrates the notion of risk. Thus, if the hazard rate is equal to 1.5 (if $h(t) = 1.5$ whatever $t \geq 0$), it means that 1.5 is the average expected number of events in a time unit, and, conversely, $1/h(t)$ measures the average expected duration before an occurrence, here 0.667 of a time unit.

---

[2] Here we understand the potential uncertainty. As will be seen later, we often chose the continuous case because of the obvious criteria of the facility of calculation.

Lastly, concerning the choice between discrete time and continuous time, both calculation methods give very similar results in the analysis of demographic phenomena. Consequently, the choice of the method depends rather on the calculation costs.

This alternative for the treatment of data in discrete time is proposed by some authors and developed mainly by biometricians. Thus, the life of each individual may be considered as a succession of Bernouilli trials,[3] occurring in each time interval until the event under study takes place. Hence, a life-course can be summed up as a succession of 0s as long as the individual stays in the state preceding the event, and ending with a 1 for the interval in which the event occurs. This kind of approach can be easily used with small samples where the observation period is not too long, nor subdivided into too small intervals (for the data-processing cost grows with the number of cases in the envisaged table).

We must insist, however, on the fact that these methods will almost always lead to results very similar to those obtained through methods in continuous time (Arjas and Kanjas, 1992). In fact, as soon as the intervals taken into account diminish, the models in discrete time converge towards the equations of the models in continuous time (Allison, 1982).

2.2. Analysis of a Heterogeneous Cohort and of the Interaction between Phenomena

There is, *a priori*, no reason why the hypotheses underlying the previous model should be verified. In the first place, the real cohorts are not formed by identical individuals, and it is unlikely that their behaviour patterns should be the same. Secondly, the past of each member in the cohort undoubtedly affects his behaviour patterns. Again, the wide variety of life-courses induces different behaviour patterns.

Various methods have been proposed to take account of this heterogeneity. We briefly present them here, to show why their hypotheses are too restrictive, before developing a more general model, which is unfortunately too complex to be entirely estimated. In this book,

---

[3] X is a variable of which the distribution follows a Bernouilli law. Let p be a value contained between 0 and 1: then $P(X=0) = 1-p$; $P(X=1) = p$.

Formalization of the Analysis

therefore we must limit ourselves to approaches that are more partial, but which our data permit us to estimate.

2.2.1. Differential demography

We can call differential demography 'the study of the differences between the various categories (ethnic, religious, social, etc.) of the population' (Henry, 1959). It therefore aims at breaking up the whole population observed into sub-populations (men and women, for instance, or individuals classified by educational status). Applying the classic demographic analysis to each sub-population makes it possible to compare specific behaviour patterns. This analysis is essentially cross-sectional.

What is of interest to us here is the use of these methods in longitudinal analysis. This use is possible if we introduce a heterogeneity in the observed cohort and if its effect on hazard rates at each age can be ascertained. Thus, for instance, we can compute the nuptiality rates of the women in a cohort according to their educational status. Longitudinal comparison is then of great interest.

This comparison, however, suffers from some limitations, which can lead to erroneous results if they are not allowed for.

In cross-sectional analysis, the individual's category at the time of the observation is taken into account. The analysis of the differences in nuptiality, fertility, or mortality can be made on a wide range of categories, social status categories, or categories defined in relation to the tenancy status of the dwelling, etc., at the time of the observation. On the other hand, longitudinal analysis requires categories that are defined once and for all and are not liable to change over the individual's life-course. While the nuptiality of a cohort of men and women can be compared, it is more delicate to compare the nuptiality of individuals according to their educational status, for some individuals marry before ending their studies. But it is no longer possible to highlight the impact of an individual's occupation on her nuptiality or fertility, since her occupation can undergo a considerable change over the life-course.

We could be induced to solve this problem by defining sub-populations by a *posteriori* criteria. Thus, we shall study the nuptiality of individuals in relation to their educational status at age 50. If the cohort is observed until that age, it is possible to define various sub-populations and to carry out a study of their previous nuptiality in a classic manner. This study does not seem to us very satisfactory, for it defines sub-populations in a set and definitive way. Rather than the educational status at age 50, it is the course of the schooling, the university studies, or even studies undertaken simultaneously with a job that influence the probability that a given individual will marry. Hence it is not through a differential approach that this question can be answered. Instead, we need an approach that follows the individual over her entire life-course in various fields. The same is true if we want to analyse the impact of the changes of occupation of an individual on her nuptiality or fertility or the impact of any other characteristic not set in time.

Even in the case where the population can be differentiated according to well defined or stable characteristics (social background, size of the family of origin, rank of birth of the respondent, etc.), its breakdown into sub-populations can quickly lead to too few individuals being at risk. No significant differences will then appear, although the tests are completely feasible (cf. Section 4.3 below).

We shall see later how using more formalized models (Markov models or proportional hazard models, for instance) makes it possible to avoid breaking down the population and provides more satisfactory tests, even when considering many characteristics.

2.2.2. Markov model or semi-Markov model

The Markov models constitute a first step in the simultaneous analysis of transitions between a certain number of states. They have long since been applied to demographic phenomena; e.g. the study of mortality by cause (Keyfitz, 1968), and the study of the demograhic evolution of several regions between which internal migrations occur (Rogers, 1973a, b). Here we shall briefly present their underlying hypotheses, as well as the main results to which they lead.

The *Markov processes*, which are used in demography, are developed in a finite state space. The moves of individuals between these different states occur within a time-frame corresponding to the age or the duration since a reference event. Thus, this event might be marriage for a study of the couple's migrations. The states of the space are identified by the individual's status. For instance, if we study the migrations of French married couples, the various states are their regions of residence over their whole marital life. Lastly, the transitions are characterized by the passage from one state to another. In our example, these transitions are the different migrations between the French regions.

The studied phenomena are thus characterized by the intensity of the transitions from one state to another (or to others).

In the most simple situation there are only two states; 'alive' and 'dead', the latter being obviously absorbing. But most steps in the life-cycle are transient states with a transition always possible towards the absorbing state (death), so that the model can easily become complicated.

Note also that, in studying human phenomena, the Markov chain which is used to modelize them can often be considered hierarchical, in the sense that, once you leave a state, you cannot come back to it. (After the birth of the first child, only a second child can be born.) The states of this particular chain are ranked hierarchically towards the final state, each state being transient with no possibility of going back.

Before formalizing the model, let us examine in more detail its underlying hypotheses. This process can involve a great number of states in which the same individual can sojourn at various moments in his lifetime. In the case of interregional migrations in France, for example, the country breaks down into 21 regions, between which we can estimate 420 migration streams. Hence individuals can follow very complex paths. The conditions under which they are followed if the model is verified, however, are oversimplifying. The probability of migration from one region to another depends on the individual's age. But it is independent of the time spent in the region of origin, of the various regions previously visited, and of the time spent in those regions. Thus, the probability of a return migration to a region is the same as a first migration to it.

Through this example, we understand better how unlikely it is that these conditions can be verified. Such a Markov model provides a simplified and very inaccurate description of the reality. It is then necessary to remove some of its conditions to get nearer to a realistic model, but this complicates to a great extent the formulation of the model (Courgeau, 1987*c*).

Note also that these models have often been applied to exhaustive populations. Under such conditions, the computations of rates and probabilities lead to results that can be easily interpreted for they have a very small variance. It becomes necessary to compute this variance when the sample decreases, in order to compare various estimations of transition probabilities (Hoem and Funck Jensen, 1982).

Formalization of the model

As shown previously, time must be marked on a common scale which we shall call the survival time or duration of stay (which can either be the time elapsed since an initial event or the calendar age). We have a finite state-space. The transitions from one state to another may be duration-specific, but will be independent of the previous steps and of the durations of stay in these steps.

Let

- t and t' be durations of stay;
- I be the finite state-space;
- $I_i(\cdot)$ be the indicators such as
  $I_i(x) = 1$ if the individual is in state i at time x;
  $I_i(x) = 0$ if not.

$P_{ij}(t, t') = P(I_j(t') = 1 \mid I_i(t) = 1)$ is the Markov transition probability from state i to state j between t and t'. The intensity of the transitions between i and j is measured by the instantaneous failure rate, which retains the same meaning as in the beginning of the chapter:

$$h_{ij}(t) = \lim_{t' \to t} P_{ij}(t, t') \frac{1}{t' - t}, \qquad (27)$$

which exists whatever $i \neq j$.

We also define

$$h_i(t) = \lim_{t' \to t} \left(1 - P_{ii}(t, t') \frac{1}{t' - t}\right) = \sum_{j \neq i} h_{ij}(t). \qquad (28)$$

In order that $\Sigma_j h_{ij}(t) = 0$, let us assume that $h_{ii}(t) = - h_i(t)$.

The probabilities of transition verify Kolgomorov's differential equations; that is,

$$\frac{dP_{ij}(t, t')}{dt'} = - P_{ij}(t, t') h_j(t') + \sum_{k \neq i} P_{ik}(t, t') h_{kj}(t'), \qquad (29)$$

which also implies that $S_i(t)$, the survivor function in i until time t is

$$\frac{dS_i(t)}{dt} = - h_i(t) S_i(t), \qquad (30)$$

Formalization of the Analysis

which can then be solved:

$$S_i(t) = \exp - \int_0^t h_i(x)dx. \qquad (31)$$

Under the hypothesis—which will always remain ours—of constant intensities by interval, the instantaneous failure rates are estimated by the hazard rates, which for each interval of time indexed by k are equal to $q_k = n_k/T_k$, where n_k is the number of transitions observed, and T_k is the total duration observed in the given state. If the size of the sample N grows, then $T_k/N \to \tau_k > 0$ for all time intervals and the $U_k = \sqrt{N}(q_k - h_k)$ are independent, with a normal distribution, with a mean of 0 and a variance $\sigma_k^2 = q_k/\tau_k$ estimated by

$$\hat{\sigma}_k^2 = Nq_k/T_k.$$

It is simple to calculate the variance when the size of the sample is not too small.

Relation between the model and the Poisson processes

Let us consider here a homogeneous Markov chain in a finite state-space of dimension r, of which the transition probabilities verify $dP(t)/dt = QP(t)$ and $P(0) = I$ identity matrix, of which the only solution is given by

$$P(t) = e^{tQ}, \qquad t > 0, \qquad (32)$$

where Q is a $r \times r$ matrix of which the independent components verify

$$-\infty < q_{ii} < 0, \qquad q_{ij} \geq 0,$$

for $i \neq j$ and $\Sigma_{j=1}^r q_{ij} = 0$. The substantive interpretation as follows.

Matrix Q:

- $-q_{ii}\,dt$ is the instantaneous probability that an individual in state i at time t exits from that state during $[t, t+dt[$.
- $q_{ij}\,dt$ is the probability that an individual in state i at time t moves to state j during $[t, t+dt[$.

Matrix $P(t)$:

- $p_{ij}(t)$ is the probability that an individual leaving i might be in state j after a duration of stay equal to t.

As shown above, we directly estimate the q_{ij} from the observed transitions as instantaneous failure rates, and

$$\hat{q}_{ii} = - \sum_{j \neq i} \hat{q}_{ij}.$$

To develop such a mobility model for a heterogeneous population, we consider the Q matrices of the particular from $Q = \lambda(M - I)$, and the transition probabilities are then governed by

$$P(t) = e^{tQ} = e^{t\lambda(M - I)}. \tag{33}$$

Let us assume that the survival times are distributed exponentially with parameter λ (which means that events occur according to a Poisson process), and that the average survival time in a state is constant (equal to $1/\lambda$). Then an individual in state i stays in it for a period τ_0, the duration of which is distributed exponentially according to

$$\text{prob}(\tau_0 \geq t) = e^{-\lambda t}, \qquad t > 0.$$

At the end of the period, he moves into j with a probability m_{ij}. (This time we do not assume that $m_{ii} = 0$; thus, an individual may stay in the same state.) Then he stays in j for a period $\tau_1 \ldots$

In that case, the transition probabilities from state i to state j are independent of the date and history of the process the individual has gone through. We then have

$$-q_{ii} \, dt = -\lambda(m_{ii} - 1) dt,$$

as the probability of leaving state i during $(t, t + dt)$ and

$$q_{ij} \, dt = \lambda m_{ij} \, dt$$

as the probability of going from i to j during $(t, t + dt)$. An interpretation of the model refers to the subordination of a process $Y(t)$ to a Markov process $X(t)$ using a Poisson process $T_\lambda(t)$ as operational or intrinsic clock (Feller, 1968).

If $(Y(t))$, $t > 0$, are the random variables that describe the biography of an individual, then they are written

$$Y(t) = X(T_\lambda(t))$$

where X is a discrete Markov chain with M as transition matrix and

$$\text{prob}(Y(t) = j \mid Y(0) = i) = p_{ij}(t)$$

is the (i, j) component of $P(t)$.

Formalization of the Analysis

The equation can then be interpreted thus:

1. Wait in a state i until the first jump time of a Poisson process.
2. At this point change state once (or remain) according to the laws of a discrete Markov chain of which one-step transition matrix is M.
3. Wait in the new state j until the Poisson process jumps for a second time....

Hence the probability that exactly n transitions happen during $(0, t)$ is

$$\text{prob}(T_\lambda(t) = n) = \frac{(\lambda t)^n e^{\lambda t}}{n!}. \tag{34}$$

In addition, the destinations of the transitions are governed by

$$\text{prob}(X(k+1) = j \mid X(k) = i) = m_{ij},$$

so that this matrix M provides a static picture of the population at an instant of movement, and the dynamics are regulated by the intrinsic clock $T_\lambda(t)$. Thus, we have

$$P(t) = e^{t\lambda(M-I)} = \sum_{n=0}^{\infty} e^{-\lambda t} \frac{(\lambda t)^n}{n!} M^n. \tag{35}$$

This general term makes it possible to capture a construction of the model according to which the propensity to change and the probability of the destinations are identified separately by λ and M.

Generalization of the model
In spite of its rather complex formulation, we have indicated the extent to which this model is a caricature of human behaviours. In fact, while it seems satisfying to take account of the durations of stay from the beginning of the observation period in modelling one phenomenon only, it is necessary to go further when modelling a large number of phenomena.

Thus, it is an acknowledged fact that, the longer the time spent in a given state (residence, professional occupation), the more the propensity to leave it decreases. In the study of migrations, many papers have presented stochastic models of spatial mobility (Courgeau, 1973), of which the cumulative inertia law is a central notion (McGinnis, 1968); and findings on the semi-Markov processes have been revived in the field of migrations (Ginsberg, 1971), as well as labour mobility (Singer and Spillerman, 1974). The presence of this cumulative inertia no longer fits the Markov model, in that it introduces, besides the life-span of the individual, the duration of stay in each of the successive states. The process thereby becomes a non-Markov process. In fact, while it is possible to take account of the time elapsed since the origin in a modelization then called a non-homogeneous Markov chain, taking account of the time spent in the last state cannot easily be incorporated.

To remain in a Markov process, one of the solutions adopted consists in demultiplying the state-space. In fact, to solve the problem of the effects of age or of generation which arises in the analysis of a sample of individuals, we can divide the sample so that the process occurs in a new state-space which makes it Markovian. The transition probabilities are then equally specific to the inertia of each considered state.

It is also possible to develop a non-Markov model which introduces an unobserved heterogeneity in the population.

Formulation with a Poisson process is very useful when modelizing the heterogeneity within a population (Singer and Spillerman, 1974). Instead of a unique coefficient λ, a matrix of coefficients λ_i can now be envisaged, according to the individual's initial state.

Assuming that the influence of unobserved factors (propensity of individuals to experience the event) can be summarized in a vector U of components $\{u_1, \ldots, u_k\}$ of which the associated probabilities are $\{\lambda_1, \ldots, \lambda_k\}$, and that the λ_i are independent of the observed individual characteristics, Heckman and Singer propose to consider a proportional influence of the unobserved factors on the survivor function. Thus, to compute the contribution of an observation to the likelihood, we sum on the unobserved:

$$S(t) = \Sigma \, S(t \mid u_i) \lambda_i \tag{36}$$

is the contribution of an observation censored at time t.

The simplest among the models will consider only two points, each individual being either of one type or the other, indexed by $u_1 = 0$ and $u_2 = \theta$, of which the associated probabilities are λ and $1 - \lambda$.

Thus, the modelization of heterogeneity brings together two distributions, one of which might be discrete. The survivor function at time t is then

$$S(t) = \lambda S(t \mid 0) + (1 - \lambda) S(t \mid \theta). \tag{37}$$

Formalization of the Analysis

The mover–stayer model is an example of this model (Section 7.1.2).

In that case, one of the groups is considered as never being at risk, so that the survivor function becomes

$$S(t) = \lambda S(t \mid 0) + (1 - \lambda) \tag{38}$$

and the hazard rates are

$$h(t) = f(t \mid 0) / [\lambda S(t \mid 0) + (1 - \lambda)]. \tag{39}$$

This model has many applications in demography, among which is the study of the first marriage carried out by Coale and McNeil (1972). In their analysis, the λ parameter measures the proportion of the population at risk, and the survivor function in the single state $S(t \mid 0)$ for those who are likely to marry is constructed as the *convolution*[4] of a normal distribution and of 'delays' distributed exponentially. We shall develop this model in more detail in Section 7.1.2, introducing a more complex unobserved heterogeneity.

As a conclusion, we can say that the introduction of Markov models has constituted a very fruitful breakthrough in the formulation of complex demographic situations, involving many states. Their aim is essentially descriptive, which is why they are used mainly for population projections. On the other hand, they do not allow a more exhaustive analysis and explanation of human behaviour patterns. It is that type of analysis which we shall now present.

2.3. Towards a More Exhaustive Analysis of Human Behaviour Patterns

With regard to the Markov models, the purpose of the analysis has changed. We concern ourselves not with forecasting the populations in the various states over time, but with trying to determine what previous steps could have led an individual to his actual state.

In order to do that, we shall consider a biography as the result of a complex stochastic process, which is influenced by various characteristics of the individual's environment.

---

[4] T and U being two independent random variables, the law of their sum is $P_{U+T} = P_U + P_T$; that is, if their densities are respectively f_U and f_T, the density of their sum is constructed as the *convolution* of the densities:

$$f(t) = f_U * f_T(t) = \int f_U(t-v) f_T(v) \, dv.$$

In the course of time, the individual successively experiences various events which can be of different types, and is subjected to varying external conditions. Thus, for example, the individual may start by experiencing a migration before assuming a first job, followed by a marriage and a simultaneous second migration; then he has successively two children, before undergoing a third migration; etc. Meanwhile, the individual may also be subjected to extremely strong external conditions on the part of his parents (careers and schooling advice related to parental prejudices, for example); then, after he has started to work, this parental influence may diminish and he then has to face labour market conditions which influence his career in a specific way; later, his marriage may create new conditions, relating to his job or personal relations; etc.

Hence we see that each event can be characterized by its date of occurrence and its nature (marriage, birth, migration, occupational change, etc.). We are thus able to point out the interactions between demographic phenomena and to analyse them in correct terms. Simultaneously, at each moment of the individual's life, we define a certain number of characteristics which may influence his life.

We then represent the existence of an individual by a series of couples of random variables (T_i, J_i) where T_i is the occurrence of the ith event of the J_i type and by a series of variables $z_i(t)$, representing all the individual's characteristics before time t. The various instants verify the relation

$$T_1 \leq T_2 \leq T_3 \ldots \leq T_i \leq \ldots$$

Similarly, as for a homogeneous population, we can define the hazard rate of the ith event of the J_i type, for an individual with the characteristics $z_i(t)$ at time t:

$$h_{ij}(t, z_i(t)) = \lim_{\Delta t \to 0} \frac{1}{\Delta t} P(T_i < t + \Delta t, J_i = j \mid T_i \geq t;$$

$$(t_1, j_1); \ldots ; (t_{i-1}, j_{i-1}); z_i(t)), \tag{40}$$

where t_k stands for the time of occurrence of the kth event of type k and $z_i(t)$ stands for the whole:

$$z_i(t) = \{z_k(t) : (t_k, j_k), k = 1, \ldots, i-1\}.$$

We can also compute the hazard rate for the ith event, whatever its type, in the form

Formalization of the Analysis

$$h_i(t, z_i(t)) = \sum_{j=1}^{m} h_{ij}(t, z_i(t)), \quad (41)$$

where m is the total number of event types to which the individual is subjected at time t.

When $z(t) = z$, that is, when the individual's characteristics do not depend on t, we can easily compute a likelihood function corresponding to the various events of the individual's biography.

For the first period observed $[0, t_1]$, we must introduce the probability that he has experienced no event before t_1. As before, this probability is equal to

$$\exp\left(-\int_0^{t_1} h_1(t, z)dt\right).$$

Then, there is the individual experience the event of type j_1 at time t_1, the probability of which is written

$$h_{1j_1}(t_1, z)dt.$$

From the couple (t_1, j_1), we can compute the probability of experiencing no event over the period $[t_1, t_2]$ which is equal to

$$\exp\left(-\int_{t_2}^{t_1} h_2(t, z)dt\right).$$

We can then continue this procedure until the time of the survey, contributing to the following likelihood:

$$\prod_{i=1}^{m} \left[\exp\left(-\int_{t_{i-1}}^{t_i} h_i(t, z)dt\right)\right]\left[h_{ij_i}(t_i, z)\right]^{\delta_i}, \quad (42)$$

where δ_i is a variable equal to 1 except for the end of the last open interval, where it equals 0. By making the product of all individual likelihood functions, we obtain the likelihood of the observed sample. Applying the maximum likelihood method leads to the estimation of all the parameters of the model, if there are not too many.

This method can be generalized in the case where $z(t)$ depends on the time by using integral products (Kalbfleish and Prentice, 1980: 182).

Actually, given the very large number of hazard rates to be estimated, the important number of characteristics to be considered, and the limited number of respondents, the analyst cannot yet estimate this joint distribution in its totality. He can only use partial approaches, which, under certain hypotheses, make it possible to estimate these hazard rates and the impact of some individual characteristics on them. Nevertheless, the general presentation we have made here remains valid for all the more partial analyses.

The non-parametric analysis will provide a first series of answers on the differentiation of behaviour patterns. The term 'non-parametric' refers to methods that do not presuppose the distribution of the event or events studied. No known distribution is fitted to the available data. For that reason, this first type of analysis—preliminary in the complete study of a phenomenon—is similar to the classic methods of longitudinal analysis used in demography. However, it involves the possibility of analysing several events at the same time. A second difference with the classic methods of analysis is to be found in the systematic computation of variances, standard deviations, and covariances of the estimators, and therefore of complex statistics allowing comparisons. In fact, the data available for these analyses, whether for the 'Triple Biographie' survey or for other samples, are never exhaustive data on a population, and the stratifications that are made make it necessary to calculate confidence intervals for the estimations.

Even when the data are exhaustive, the large number of situations in which an individual can be placed will reduce the numbers at risk and will imply the computation of these variances and covariances. Thus, in presence of a large sample, the very subtle stratification done on the data (by state, by age, and by the time elapsed since an initial event, by type, etc.) makes it necessary to present these computations so as to make the reasoning totally reliable.

In Part II we shall introduce some characteristics of the individuals concerned by parametric methods, which constitute a generalization of the current regression methods to biographies. Lastly, semi-parametric methods will provide a synthesis between the non-parametric and the parametric approaches.

2.4. Conclusion

In this chapter we have presented the vocabulary of the analysis of survival times (duration of stay in a demographic state), as well as the more current notations, in the simplified context of a single event, which may occur in a homogeneous cohort. We have also discussed the matter of continuous or discrete time.

To take into account the heterogeneity of the cohorts and the interaction between phenomena, we show that the methods of differential demography, useful in a transversal context, prove to be insufficient in the longitudinal context. Similarly, the Markov models, useful for making population projections, are not appropriate to the analysis and the explanation of interactions between demographic phenomena. Therefore, we present in the most general case the methods of analysis which are proposed in this book and which constitute a generalization of differential demography and of Markov models. Unfortunately, it is not yet possible to apply this overall model to survey data, for it requires the estimation of too great a number of hazard rates and parameters. Hence we shall be led to limit the field of application to a smaller number of events and individual characteristics. Nevertheless, these analyses are very complex and will enable us to do the groundwork in this area.

We must now confront the instruments with the data, beginning with the least ambitious analyses and always taking account of the limits imposed by the nature of the collected information.

3

Methods of Estimation using Censored Observations

In this chapter we shall present the principle of non-parametric estimation methods of hazard rates using various types of survey data, which can be either right-censored or left-censored.

3.1. Censoring Problems

There are different types of 'gaps' in retrospective or prospective longitudinal data, depending on the collect. If the observed time interval runs from date t to date t', and if the evolution of a process starting in t_0 and ending in t_1, with $T = t_1 - t_0$, is studied, four situations appear, excluding the case where no part of the process is observed (Fig. 3.1):

1. The entire process is observed and the interval is not censored.
2. t occurs after the start of the process, and the data are left-censored.
3. t' occurs before the end of the process, and the data are thus right-censored.
4. The data may be censored both on the left and on the right. (The terms 'left-censored' and 'right-censored' intervals are also used.)

Therefore, a simple process refers to an event that defines an interval, such as a first marriage or death. But renewable events such as births or successive migrations must also be envisaged. This second type of process (Fig. 3.2) takes place between t_0 and F, the end of the process, which may happen before the individual's death (end of the reproductive period).

It is essential to take account of incomplete observations to estimate the different hazard rates or the average waiting times in a given state, but this gives rise to difficulties. Depending on the type of censoring, different responses are proposed.

Methods of Estimation

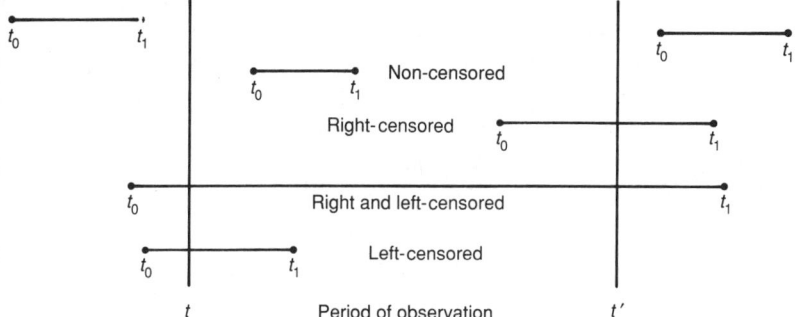

FIG. 3.1. Different types of censoring of a simple process starting in t_0 and finishing in t_1 when the observation period starts in t and finishes in t'

Here we propose two types of solution concerning right-and left-censored data.

As has already been mentioned, whatever the method of collect, observation intervals are nearly always right-censored. We shall first of all show that taking account of this type of censoring is not too difficult, and then we shall review the hypotheses underlying the applications in this case (Feller, 1968).

We shall then proceed to consider left-censored data, which are also very common in practice. They are particular in that their use biases results. So far, no efficient method has been available to counteract this bias. Solving this left-censoring is more difficult, because it is impossible to assess the effects of previous developments of the

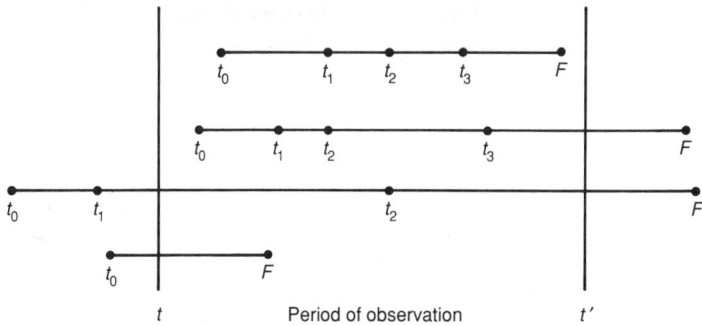

FIG. 3.2. Renewable events ending in F with the same observation period $[t, t']$

process on its observation in the present, or on its forecast. Given the absence of method, one tends to choose between two hypotheses. The first one, classical but most unsatisfactory, was that the process starts in t, even though it is known to be anterior. The second one, clearly not very realistic, was that the unknown past of the process in no way affects its present development.

3.2. Right-censoring

Estimations using retrospective data are possible by taking account of the observations right-censored by the data of the survey. Evaluation of the average waiting time between events and that of hazard rates for these events is then partly based on observation of incomplete time-spans.

However, when only the last occurrence (for instance, the last migration or the last birth, etc., before the interview) is observed, the problem can be resolved only by imposing restrictive conditions.

If we consider the case of renewable events which follow a Poisson distribution—change of domicile or job, births, etc,—then the durations of stay are distributed exponentially.

In a retrospective survey all the intervals are observed, while only the last is censored. However, sometimes only the last interval between the last event and the interview is observed in t': Y_1 (Fig. 3.3). The last event observed took place at time $t_v < t'$. The next (unobserved) will (perhaps) take place at date $t_\mu > t'$.

If the next occurrence happens in t_μ, the waiting time $t_\mu - t'$ is random, and, owing to the lack of memory of the Poisson process, Y_2 is distributed exponentially. It has moreover been demonstrated that Y_1 is also distributed exponentially.

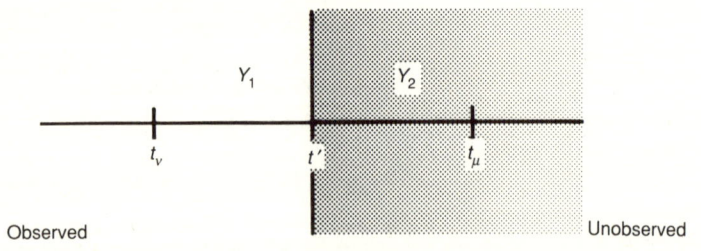

FIG. 3.3

Methods of Estimation

Because of this, the truncated intervals correspond to longer durations of stay on average than those observed in the rest of the event history. In fact, the distribution of right-censored durations of stay is not exponential, but has a gamma density. The expectancy of the duration of stay censored by the interview is double that of other inter-event intervals. This demonstrates that a random interruption of a stochastic process occurs more probably during long inter-event intervals than short ones: this corresponds to the intuitive notion that the interview is more likely to interrupt longer waiting times between events.

Thus, a systematic omission of durations of stay censored by the end of observation leads to considerable underestimation of the mean durations of stay, with a consequent overestimation of the instantaneous failure rates from one state to another. In fact, long durations of stay would be systematically discarded in this case.

Finally, it must be remembered that, to use the right-censored intervals, we have had to assume the independence of the event being studied, from the end of the observation. The individuals whose behaviour is unobserved would have behaved as those under present observation have. It is possible that this hypothesis, while realistic, will not be checked, in which case the censored data must be processed differently—by exclusion, or separate study, for instance.

Estimation using retrospective data will therefore be undertaken by taking account of the intervals censored by the date of the survey. Various estimators may be used depending on the type of data collected. Here we present the Kaplan–Meier or product-limit estimator, when only one occurrence is observed, and the Aalen estimator, which generalizes the precedent and applies it to more complex situations.

3.2.1 The Kaplan–Meier or Product-Limit Estimate

For the formalization of non-parametric estimation, the propositions of Kaplan and Meier (1958) are still a reference. We will outline their techniques before describing them in detail in order to show first of all how innovative they were, and then the additions that were made to them. These authors were the first to concern themselves with estimation of right-censored data. Also, they made it possible to estimate a survivor function which took account of this.

As seen before, these censorings are due to the nature of the data. Whatever type of collect is used, event history data are censored at the interview or definitely at death. We have at our disposal, therefore, a sample of individuals for whom, when the event studied has occurred during the observation period, this has been recorded. If this is not the case, the censoring time for various reasons such as death, migration, date of the survey, etc., is recorded, resulting in a loss of information at a given date.

The product-limit estimate can be shown to maximize the (generalized) likelihood. If occurrences are observed in $t_1 < t_2 < \ldots < t_k$, the (generalized) likelihood of the observation is formed for each age t_i by the following contributions:

$$L_i = h_i^{d_i}(1 - h_i)^{N_i - d_i}$$

(formula (18), Section 2.2), where
 d_i is the number of occurrences in t_i
 N_i is the population at risk just before t_i
 h_i is the instantaneous hazard rate in t_i
The logarithm of generalized likelihood is then

$$\log L = \sum_i [d_i \log h_i + (N_i - d_i) \log (1 - h_i)], \qquad (1)$$

and the maximum likelihood estimator is obtained by deriving the logarithm of likelihood as a solution of

$$d \log L / dh = 0.$$

That is, for each t_i,

$$\hat{h}_i = d_i / N_i. \qquad (2)$$

The variance of this estimator is calculated using the standard theory of large numbers—which may limit its use when samples are found to be too small.

Asymptotically, $\sqrt{n}(\hat{h}_i - h_i)$ will have a multivariate normal distribution with mean zero and a convariance matrix which is estimated by the inverse Fisher information matrix, thus:

$$\frac{d^2 \log L}{dh_j \, dh_k} = \begin{cases} -\dfrac{N_j}{\hat{h}_j(1 - \hat{h}_j)} & \text{if } j = k, \\ 0 & \text{if } j \neq k, \end{cases}$$

so that

Methods of Estimation

$$\text{var}(\hat{h}_j) = \frac{d_j(N_j - d_j)}{N_j^3}. \quad (3)$$

The same result would, indeed, be obtained for independent binominals.

The estimator of the survivor function, called the Kaplan–Meier estimator, is obtained thus:

$$\hat{S}(t) = \prod_{t_i < t} (1 - \hat{h}_i) = \prod_{t_i < t} (N_i - d_i) N_i^{-1}. \quad (4)$$

This is the expression of the estimator when several events occur at the same time.

If the occurrences are reported exactly, one per date ($d_i = 1$), the estimator of the survivor function is defined thus:

$$\hat{S}(t) = \prod_r [(N - r)/(N - r + 1)], \quad (5)$$

where
N is the size of the sample
$t_1 \leq \ldots \leq t_N$ is the ordered ages of loss or occurrence
r is the number of losses or occurrences at age $t_r \leq t$

If no information is lost, the estimator $S(t)$ becomes the usual binominal estimator: the proportion of survivors. This new estimator is consistent and only slightly biased, and an asymptotic expression of its variance may be obtained, known as the Greenwood formula.

So, as

$$\log \hat{S}(t) = \sum_{t_j} \log(1 - \hat{h}_j),$$

the asymptotic variance of $S(t)$ is estimated thus:

$$\text{var}\{\log(\hat{S}(t))\} \simeq \sum_{t_j} \text{var}[\log(1 - \hat{h}_j)]$$

$$\simeq \sum_{t_j} \left(\frac{1}{1 - \hat{h}_j}\right)^2 \text{var}(\hat{h}_j)$$

$$\simeq \sum_{t_j} \left(\frac{1}{1 - \hat{h}_j}\right)^2 \frac{\hat{h}_j(1 - \hat{h}_j)}{N_j}$$

56   Extending the Scope of Longitudinal Analysis

$$\simeq \sum_{t_j} \frac{d_j}{N_j(N_j - d_j)}.$$

Hence

$$\text{var}\{\hat{S}(t)\} = (\hat{S}(t))^2 \sum_{t_j} \frac{d_j}{N_j(N_j - d_j)}, \qquad (6)$$

which is known as Greenwood formula.

This variance may be used to compare different populations and was first derived as the asymptotic variance of a classical life table estimator.

Because of the conditions on the censoring, the confidence interval for the survivor function can be calculated for extreme values of t. But such an approximate confidence interval may include impossible values outside the range [0, 1]. This problem can be avoided by applying the asymptotic normal distribution to $S(t)$, which is not confined to this range. Kalbfleisch and Prentice suggest, in this case, the calculation of the asymptotic variance s^2 of

$$\log(-\log S(t)) = SA(t)$$

estimated by

$$\hat{s}^2(t) = \sum_{i \mid t_i < t} \frac{d_i}{N_i(N_i - di)} \Big/ \left[\sum_{i \mid t_i < t} \log\left(\frac{N_i - d_i}{N_i}\right) \right]^2. \qquad (7)$$

A confidence interval at 95 per cent for $SA(t)$ is given by $SA(t) \pm 1.96\, s(t)$, which corresponds to a confidence interval for $S(t)$ of

$$S(t) \exp(\pm 1.96\, s(t)),$$

which then takes its values in [0, 1].

However, if the censoring is considered discrete, and the distributions continuous, other estimations of the instantaneous hazard rate are possible.

Examples
1. If the events are reported exactly and there are no simultaneous occurrences, the second formulation of $S(t)$ may be applied.

Let there be 10 individuals. Deaths are observed at 0.8, 1.0, 7.6, and 9.2 years, and losses at 0.4, 2.1, 4.7, 5.3, 8.6, and 14.0 years.

Methods of Estimation

$$\hat{S}(t_1) = S(0.8) = \frac{8}{9} = 0.888$$

$$\hat{S}(t_2) = S(1.0) = \frac{7}{8} \times \frac{8}{9} = 0.777$$

$$\hat{S}(t_3) = S(7.6) = \frac{3}{4} \times \frac{7}{8} \times \frac{8}{9} = 0.583$$

$$\hat{S}(t_4) = S(9.2) = \frac{1}{2} \times \frac{3}{4} \times \frac{7}{8} \times \frac{8}{9} = 0.292$$

2. Let us now examine the case where several individuals experience the event studied at the same time.

Let there be a sample formed by the 250 inmates of a young people's home; the event studied is finding a job. The individuals are studied while they remain in the home. For those who leave before finding a job, only the date of their departure from the home is known. The first events observed occur in the following manner, following the hypothesis that the departures occur once a job has been found:

After 6 months	31 jobs	12 departures
After 7 months	11 jobs	
After 9 months		10 departures
After 10 months	15 jobs	21 departures
		etc.

It may therefore be calculated that

$$\hat{S}(t_1) = 1 - \frac{31}{250} = 0.876$$

$$\hat{S}(t_2) = \left(1 - \frac{31}{250}\right)\left(1 - \frac{11}{207}\right) = 0.829$$

$$\hat{S}(t_3) = \left(1 - \frac{31}{250}\right)\left(1 - \frac{11}{207}\right)\left(1 - \frac{0}{196}\right) = \hat{S}(t_2) = 0.829$$

$$\hat{S}(t_4) = \left(1 - \frac{31}{250}\right)\left(1 - \frac{11}{207}\right)\left(1 - \frac{15}{186}\right) = 0.762$$

etc.

3. Finally, using a simple example, the estimations obtained may be compared by taking account of a sample as a whole, or by estimating the survivor functions separately for two sub-populations.

We can then assess the improvement brought by the estimator currently used in demography, as follows.

Let there be a sample of 18-year-old students observed over two years, the number of whom grows by 400 from the same population during the second year, and whose entries into the labour force is observed in Fig. 3.4.

The estimators of $S(1)$ calculated for each sub-population, respectively 0.45 and 0.40, with a standard deviation of 0.02, agree with the hypothesis of homogeneity of the whole sample, in which $\hat{S}(1) = 0.417$. $S(2)$ may then be calculated using only the data from the first group, that is $\hat{S}_r(2) = 60/200 = 0.3$, this being the reduced estimate of the survivor function in the student state. This is possible only when, as is assumed here, the observation limits (two or one year) are known for all students, whether they are observed to enter the labour force or not.

But a different method could have been used. Let us calculate the probability of staying for two years in the home, supposing that no job was found during the first year:

$$\hat{S}(2)/\hat{S}(1) = 60/90 = 0.667.$$

We thus compute

$$\hat{S}(2) = [\hat{S}(2) \mid \hat{S}(1)]/\hat{S}(1) = 0.667 \times 0.417 = 0.278.$$

The advantage of this calculation is that the same approach is applied to the two sub-groups irrespective of the different observation. Also,

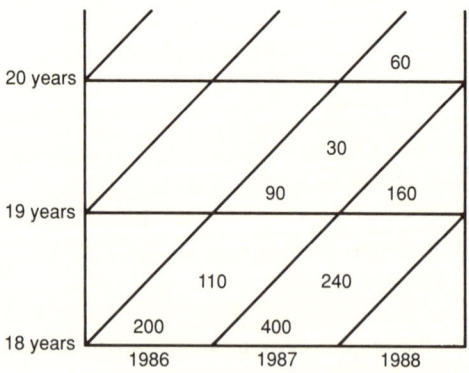

FIG. 3.4. Lexis diagram

it gives an estimation of $S(2)$ which takes account of the sample as a whole, and not just a part of it.

In fact, to calculate $\hat{S}(2)/\hat{S}(1)$, only those who are still at risk after one year of observation need to be counted, and we need to calculate $\hat{S}(1)$ of those remaining after one year; i.e., 250 out of an initial sample of 600 are counted.

3.2.2. The Aalen Estimator

Later, Aalen (1978) formalized the analysis, by generalizing the theoretical framework presented by Kaplan and Meier (1958) and Nelson (1969).

We shall here present his approach, which covers the estimations proposed by Kaplan and Meier. One of Aalen's main innovations is the possibility to take into account, when analysing a single event, several different types, such as several causes of death, and to compare their intensity within a theoretical framework not requiring a hypothesis concerning the independence of the various risks.

It is true that, if the sample is examined for mortality, without distinguishing the causes, Kaplan and Meier's estimation of the survivor function is recommended if the sample is small. But when the effects of the various causes of death must be separated, the problem of their dependence arises. If the Kaplan–Meier estimators are calculated for each cause, thereby considering the other causes of death as losses, then a hypothetic mortality due to a specific cause, not taking account of other risks, is estimated. The hypothesis of independent risks, which is obligatory in this case, is not only very strong but, above all, unverified; another estimation is therefore desirable. The problem could then be reformulated in terms of counting processes. These processes make it possible to study independent individuals at risk from various occurrences. The process $N(t)$ counts events occurring during $[0,t[$; it is univariate if there is only one phenomenon, and multivariate $(N_i(t)\ i = 1, \ldots, k)$ if it is a collection of k counting processes which may be dependent.

For each of the stochastic processes $N_i(t)$, let us define an intensity process $\Lambda_i(t)$ as the conditional probability of the occurrence of the event i during $(t, t + \Delta t)$, knowing the previous history of process $F(t)$. Then

$$\Lambda_i(t) = \lim_{\Delta t \to 0} \frac{E[N_i(t + \Delta t) - N_i(t) \mid F(t)]}{\Delta t}. \qquad (8)$$

Aalen (1978) thus introduces the multiplicative intensity model, noting the intensity process in the following form:

$$\Lambda_i(t) = h_i(t) Y_i(t), \qquad (9)$$

where $Y_i(t)$ represents, say, the population at risk of experiencing event i, and $h_i(t)$ represents the intensity of event i for an individual. (In this case, it is a classical hazard rate.) It must be noted here that $Y_i(t)$ may be defined in a more complex manner, for instance in studying the spread of an epidemic, when $Y_i(t)$ may be considered as the product of the number of individuals affected by the population at risk. The intensity remains well defined and represents contagion. The cumulative intensity is therefore denoted by formula (22) of Section 2.1.3:

$$H_i(t) = \int_0^t h_i(s) ds.$$

This cumulative intensity may easily be calculated. In the case of a cause-specific mortality hazard, for example, let $Y(t)$ be the population at risk just before instant t. Let $t_{i1} < t_{i2} < \ldots$ denote the dates of death observed for cause i. Then Nelson's estimator of cumulative intensity at instant t may be written thus:

$$\hat{H}_i(t) = \sum_{t_{ij} \leq t} \frac{1}{Y(t_{ij})}, \qquad (10)$$

and the estimator of its variance,

$$\operatorname{var}(\hat{H}_i(t)) = \sum_{t_{ij} \leq t} \frac{1}{[Y(t_{ij})]^2}. \qquad (11)$$

Naturally, several deaths by cause i may be observed at instant T_{ij}, N_{ij}, which then replace the numerator of $\hat{H}_i(t)$ and $\operatorname{var}(\hat{H}_i(t))$.

Intensities $h_i(t)$ remain well defined even when risks are dependent, and processes $\hat{H}_i(t)$ can be considered as approximately independent.

Aalen's contribution consists mainly in his application of the theory of martingales to the counting processes. The study of multivariate counting processes, the statistical theory of which has been developed in his work, has demonstrated that using the martingales theory and stochastic integrals is necessary in order to establish the properties of estimators and test statistics.

Because of this, very weak assumptions are imposed by this model framework on the N_i counting process, which also means that there is no restriction on interdependence between the $Y_i(t)$, nor on their dependence on the past.

Thus, for each t the only important restriction is that the $Y_i(t)$ must be dependent on what occurred before t (with the possibility of depending on exogenous random elements), but must in no case depend on the future of the process.

Within the framework of a non-parametric estimation of the $h(t)$, this means that the number of individuals at risk can be modified at any time t on the basis of past experience in almost arbitrary ways. Losses of information can then be taken into account.

It is thus easy to judge the far more general characteristics of these models, of which the Kaplan–Meier estimations are but a particular case. We shall detail the applications in Chapter 4.

3.3. Left-Censoring

Many examples of left-censoring are available: the place of residence of an individual at time t (start of observation) is known, but not the length of time he has resided there; the marital status at t is known, but not the date of marriage if it was before t; and so on. In the same way, if migrations of an individual are observed during interval $[t, t']$, sometimes the rank of the migrations observed is available, but in the most extreme cases no migrations are registered during observation. It has been shown, for instance, that the frequency of migrations and the distances covered during childhood constitute factors that have very considerable influence on the migratory behaviour pattern of future adults (Courgeau, 1985 b; Courgeau et al., 1986). Thus, it is very clear that the hypothesis which consists in ignoring the influence of the past on the future development of a process leads to errors of appreciation or interpretation of the phenomena observed in the case of migrations.

To test the biases introduced by left-censoring, we have the data from the 'Triple Biographie' survey, which we had artificially censored on the left. Thus, we were to test the validity of the estimation method by comparing the results obtained from one interval giving complete observation $[t, t']$ with those from intervals arbitrarily truncated on the left $[t^*, t']$.

The main objective of this method, proposed by B. Turnbull (1974), is to estimate the number of individuals who experienced the event before the start of observation (t) from

1. the number of individuals for whom the information is censored; and
2. the values of the survivor function, which is iteratively generated from initial data until a suitable convergence is obtained.

The models proposed give estimators of the survivor function. This is not restrictive, in so far as the other estimators (hazard rates, mean durations of stay) are obtained together.

3.3.1. *A method of rectification*

Let us suppose that the process studied is the birth of the first child to women in a given group of generations. Naturally, the method may be generalized to any other phenomenon studied.

Let T be the random variable representing the age of a woman giving birth for the first time, counted from the moment of her entry into the fertile population. Because only period $[t^*, t']$ is observed, some births may occur before t^*. However, at the start of observation t^*, the numbers of women who have already had their first child and those who have not are known, by generation.

We shall thus represent as

μ_j the number of mothers of age a_j for whom the date of birth of their first child is not available (left-censored observations);

r_j the number of women observed who become mothers between ages a_{j-1} and a_j;

λ_j the number of childless women whose censoring time is age a_j.

The method presented here will operate by successive iterations.

1. These iterations are begun by estimating the initial survivor functions, i.e. the likelihood of remaining childless for each age, using the sample of only those women whose ages are known at the moment their first children are born:

$$\{S_0^0, S_1^0, \ldots S_m^0\},$$

where the maximum age observed during period $[t^*, t']$ is m.

Methods of Estimation

2. Using these survivor functions, the values of the following α_{ij}^0 coefficients are estimated:

$$\alpha_{ij}^0 = \frac{S_{i-1}^0 - S_i^0}{1 - S_j^0}, \qquad i \leq j, \tag{12}$$

which represent the proportion of births at age i of women having had their first children before age j; so $P(a_{i-1} < T \leq a_i \mid T \leq a_j)$, given that the behaviour pattern of the different generations is identical. An initial estimate of the number of women who bear first children between ages a_{i-1} and a_i is then

$$r_i^0 = r_i + \sum_{k=i}^{m} \mu_k \alpha_{ik}^0, \tag{13}$$

obtained by adding to the women observed who become mothers between a_{i-1} and a_i the non-observed mothers spread out according to α coefficients.

3. Using these new values, new survivor functions may be estimated by

$$S_i^1 = 1 - \frac{r_i^1}{R_i^1} \tag{14}$$

$$S_j^1 = \left(1 - \frac{r_j^1}{R_j^1}\right) S_{j-1}^1, \quad \text{where } j = 2, \ldots, m,$$

with

$$R_j^1 = \sum_{k=j}^{m} (r_k^1 + \lambda_k),$$

which is the population at risk between a_{j-1} and a_j.

It is then possible to return to step 2 with the new survivor functions S_j^1, which give new values α_{ij}^1 and r_i^1, replacing S_j^0, α_{ij}^0, and r_i^0.

The iteration is continued until the estimators S_j^l converge. This convergence may be tested by calculating, for instance, the maximum value of $|S_j^l - S_j^{l-1}|$ and then reiterating the process as long as this value is above the threshold initially set.

3.3.2. Results obtained

The estimations were made[1] on a sample of 511 women in the 'Triple Biographie' survey, born between 1926 and 1930. The interval that

[1] Cf the work of E. Rouy (1986).

corresponds to the complete observation, with no left-censoring for the study of first births, was $[t, t'] = [1942, 1965]$. We therefore applied the method successively to the following intervals $[t^*, t']$: [1943,1965], [1947,1965], [1952,1965].

Figure 3.5 shows the different curves obtained if the interval is truncated by ten years according to how censorings are taken account of. The curves that represent the likelihood of being still childless at the ages noted on the x-axis correspond to the estimations obtained respectively from the set of complete data (survivor functions really observed), from the sample reduced to the uncensored data, and finally from the results of the model (reconstituted data).

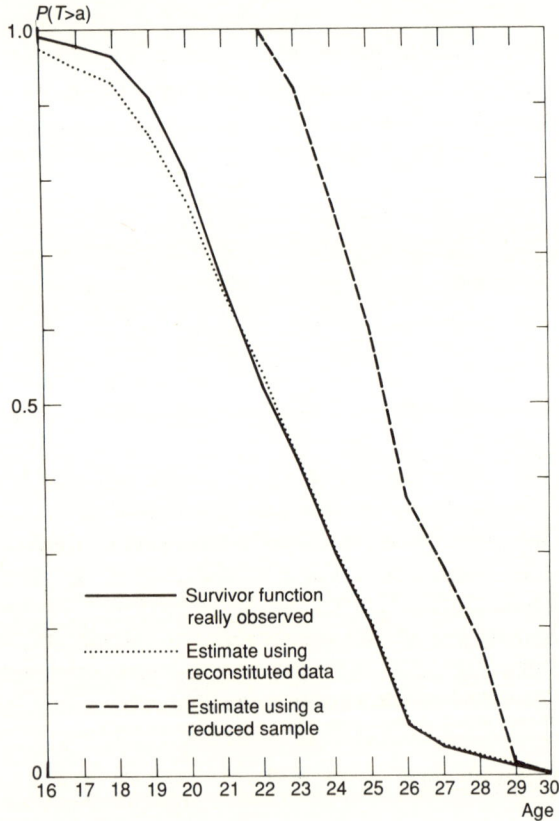

FIG. 3.5. Survivor function: age on first birth
Data left-censored by 10 years (1952, 1965)

Methods of Estimation

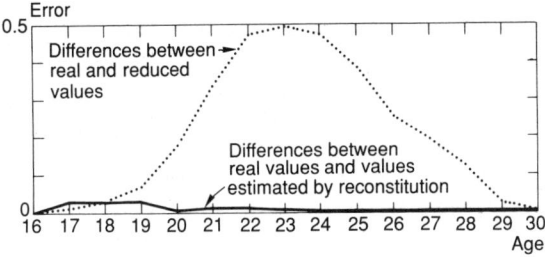

FIG. 3.6. Distribution of errors
Data left-censored by 10 years (1952, 1965)

Figure 3.6 shows the error distribution of estimated values compared with observed values for the two estimations, which are obtained by the past reconstitution model or from the reduced sample. It is clear from this that estimation by reconstitution is a good tool compared with the practice that consists in selecting only those individuals of whom observation is complete, in cases where data are partly left-censored.

However, some qualifications must be made. First, the method is based on the hypothesis that the process studied does not vary over time. Therefore, for phenomena highly influenced by their period of occurrence, or with a very pronounced evolution from one generation to another, a far less satisfactory estimation would be obtained. In this case it would be advisable to construct a term α_{ij} in the form of $\alpha_{ij}(t)$.

Moreover, this same method gives similar, although less precise, results when applied to migrations. This is because migrations are more evenly distributed over the observation interval than first births, which were concentrated to the left of the interval. To apply the model to migrations (Fig. 3.7), after considering the age of first migration (after age 14), the time interval between two successive migrations is taken into account. This index is particular in that it can range beyond the observation interval. The four possible cases are to be found in Fig. 3.2. Using these, another estimation model may be tested, one that takes account of left-censored observations in the same way as those censored on the right; but it requires more complex adjustments which we will not present here, because they are incomplete.

We have proposed the beginnings of a practice which takes account of censored data. This practice, as we have seen, does not require

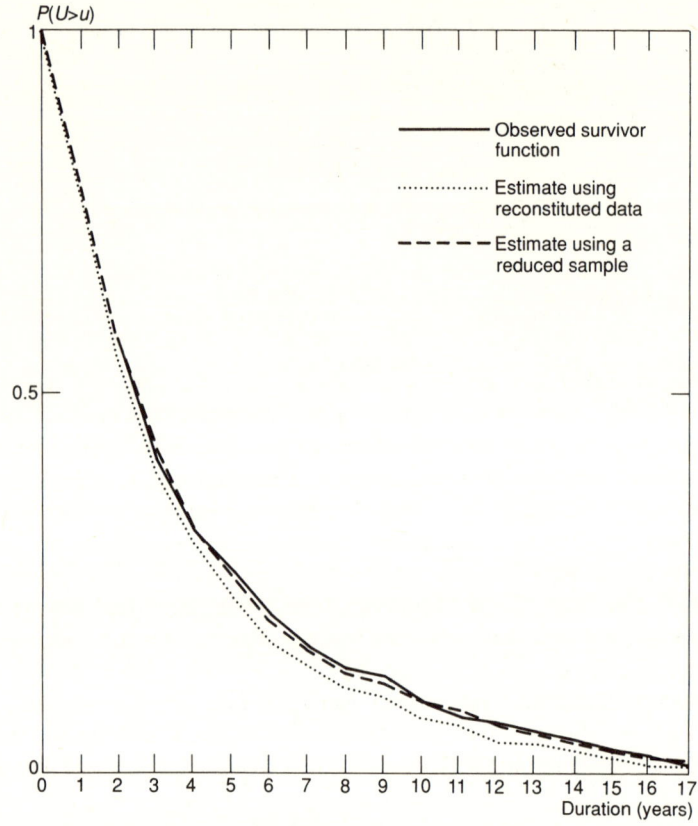

FIG. 3.7. Survivor function between first and second migration
Data left-censored by 10 years (1952, 1965)

costly statistical refinements and gives genuine improvements in the analysis. It does, however, require very strong hypotheses on the population behaviour *vis-à-vis* the phenomenon studied, which must be stationary in time for the different generations studied.

3.4. Conclusion

We have presented here two actuarial methods of non-parametric estimation from observed data. They were successively developed,

Methods of Estimation

first in response to the need for calculation using incomplete, censored information, and then to be able to take account of every fluctuation in the population at risk.

Censored data must, indeed, be taken into account if results are not to be systematically biased,

- either by underestimating all the durations of stay in right-censored cases (this possibility is now satisfactorily dealt with in all longitudinal data analysis models);
- or by limiting the study to particular sub-populations, which reduces the field of analysis;
- or, most important, by totally ignoring the past process history, which results in considerable bias, which is still not yet resolved.

The methods have proved efficient at estimating the survivor functions when there is a right-censoring. On the other hand, if the observation is left-censored, we suggest an iterative estimation model; this is, however, based on very restrictive hypotheses.

Having gone over the fundamental notions underlying event history analysis, we shall now envisage the analysis of an event in more detail, and then that of two interacting events, and finally that of several. These notions will help in describing the non-parametric estimations used.

4

Study of a Single Event

The study of a single event may be made by using a single population or by comparing several. In addition, this particular event may be of several types—migrations according to destination, for instance. We shall therefore consider the study of these different cases one after the other.

4.1. Single Sample: A Single Event

The distribution of censoring times and of events occurring in the population over time may be subjected to different hypotheses, resulting in various estimations.

4.1.1. Estimations in discrete time

Let $t_1 < t_2 < \ldots < t_k$ be the dates of successive events observed in a sample of N independent observations. Let us suppose that d_i individuals experience the event at t_i and that m_i individuals are lost between $[t_i, t_{i+1}[$. Finally, let $N_i = (d_i + m_i) + \ldots + (d_k + m_k)$ be the individuals at risk just before t_i.

The Kaplan Meier estimator, presented in Section 3.2.7 (formulas (1), (3), and (4)), applies in this case, and leads to the estimation of the hazard rate:

$$\hat{h}_i = \frac{d_i}{N_i},$$

and to the survivor function:

$$\hat{S}(t) = \prod_{t_i \leq t}\left(1 - \frac{d_i}{N_i}\right) = \prod_{t_i \leq t}(1 - \hat{h}_i).$$

The variance of $\hat{h}_i$ is asymptotically equal to

Study of a Single Event

$$\mathrm{var}\,(\hat{h}_i) = \frac{d_i(N_i - d_i)}{N_i^3},$$

and that of $\hat{S}(t)$ is given by the Greenwood formula ((6) in Chapter 3).

4.1.2. Actuarial estimation

Let us now assume that the d_i individuals experience the event in interval $[t_i, t_{i+1}[$ and that the m_i individuals are censored in the same interval. It may be assumed that the instantaneous hazard rate remains constant all through the interval $(h(t_i) = h_i)$, and that the risk of censoring is itself constant $(h_c(t_i) = c_i)$ and independent of the instantaneous hazard rate.

The contributions to total likelihood $L(h, c)$ are then as follows for each interval (conditioned by their previous stay until the beginning of the interval, with $b_i = t_i - t_{i-1}$).

1. The $N_i - d_i - m_i$ individuals who survive uncensored throughout the interval contribute by

$$S_{1i}(t) = \exp(-b_i(h_i + c_i)).$$

2. The d_i individuals who experience the event during the interval contribute by

$$S_{2i}(t) = \int_0^{b_i} h_i e^{-th_i} e^{-tc_i} dt = \frac{h_i}{h_i + c_i}\{1 - \exp[-b_i(h_i + c_i)]\}.$$

3. The m_i individuals who are censored during the interval contribute by

$$S_{3i}(t) = \int_0^{b_i} c_i e^{-th_i} e^{-tc_i} dt = \frac{c_i}{c_i + h_i}\{1 - \exp[-b_i(h_i + c_i)]\}.$$

Thus, the contribution to the log likelihood arising for the interval $[t_{i-1}, t_i[$ is

$$l_i(h_j, c_i) = -(N_i - d_i - m_i)\,b_i(h_i + c_i)$$
$$+ d_i \log\left(\frac{h_i}{h_i + c_i}\right) + m_i \log\left(\frac{c_i}{h_i + c_i}\right)$$
$$+ (d_i + m_i) \log[1 - \exp(-b_i(h_i + c_i))]. \quad (1)$$

70   *Extending the Scope of Longitudinal Analysis*

The maximum likelihood estimates h_i and c_i are then obtained by deriving the log likelihood and as a solution of

$$\frac{dl_i}{dh_i} = -(N_i - d_i - m_i)b_i + \frac{d_i}{h_i} - \frac{d_i + m_i}{h_i + c_i}$$

$$+ \frac{(d_i + m_i)b_i \exp[-b_i(h_i + c_i)]}{1 - \exp[-b_i(h_i + c_i)]} = 0,\tag{2}$$

$$\frac{dl_i}{dc_i} = -(N_i - d_i - m_i)b_i + \frac{m_i}{h_i} - \frac{d_i + m_i}{h_i + c_i}$$

$$+ \frac{(d_i + m_i)b_i \exp[-b_i(h_i + c_i)]}{1 - \exp[-b_i(h_i + c_i)]} = 0,$$

which gives, by subtraction of the two equations and then substitution,

$$\hat{h}_i = -\frac{d_i}{b_i(d_i + m_i)} \log\left(\frac{N_i - d_i - m_i}{N_i}\right) \text{ and}$$

$$\hat{c}_i = -\frac{m_i}{b_i(d_i + m_i)} \log\left(\frac{N_i - d_i - m_i}{N_i}\right).\tag{3}$$

When the interval width $[t_{i-1}, t_i]$ is small (b_i small), then $(d_i + m_i)/N_i$ is also small, and the logarithm may be expanded in series (dropping the subscripts):

$$b\hat{h} = \frac{d}{N} + \frac{1}{2}\frac{d(d+m)}{N^2} + o\left(\frac{d+m}{N}\right)^3,\tag{4}$$

where $o(\varphi(\cdot))$ indicates that $o(\varphi(\cdot))/\varphi(\cdot) \to 0$. This leads to

$$b\hat{h} = \frac{d}{N - \frac{1}{2}(d+m)}\left[\frac{N^2 - \frac{1}{4}(d+m)^2}{N^2}\right] + o\left(\frac{d+m}{N}\right)^3,$$

from which

$$b\hat{h} = \frac{d}{N - \frac{1}{2}(d+m)}\left[1 - \left(\frac{\frac{1}{2}(d+m)}{N}\right)^2 + o\left(\frac{d+m}{N}\right)^2\right]$$

$$b\hat{h} = \frac{d}{N - \frac{1}{2}(d+m)}\left[1 + o\left(\frac{d+m}{N}\right)^2\right].\tag{5}$$

Study of a Single Event

The instantaneous hazard rate estimator in interval $[t_{i-1}, t_i[$ is therefore equal to

$$\hat{h}_i = \frac{d_i}{N_i - \frac{1}{2}(d_i + m_i)} \quad \text{when } b_i = 1, \tag{6}$$

assuming that events and censoring times occur uniformly and independently of each other during the interval.

In classical demography, an annual rate like the probability of experiencing the event in the interval, without censoring, is usually calculated to be

$$\hat{h}'_i = \frac{d_i}{N_i - \frac{1}{2} m_i}. \tag{7}$$

Therefore the classical hypothesis of independence is made: those who were censored between t_{i-1} and t_i before experiencing the event would have experienced this event if they had remained, as did those observed during this period.

The two preceding formulas show what differentiates an instantaneous hazard rate from an annual one.

The graphs of instantaneous hazard rates, or more precisely of the conditional density, are often used to decide how to fit the distribution observed to a known distribution family. This will be dealt with in detail in Chapter 7.

The estimator of the probability of staying in the initial state is then equal to

$$\hat{S}(t_i) = \prod_{k \leq i} (1 - \hat{h}'_i), \tag{8}$$

and its variance is obtained in the same way as in the previous case by the Greenwood formula:

$$\text{var}(\hat{S}(t_i)) = (\hat{S}(t_i))^2 \sum_{k=1}^{i} \frac{d_k}{(N_k - \frac{1}{2} m_k)(N_k - \frac{1}{2} m_k - d_k)}. \tag{9}$$

4.1.3. Estimation of cumulative hazard rates

Another possible technique is to use cumulative hazard rates, with which the quality of fit may be better evaluated. These cumulative hazard rates are estimated by using expression (13) of Chapter 2:

$$\hat{H}(t) = \sum h_i.$$

This formula is approximative and is better verified when the values of h_i are small. Since $(S(t) = \exp(-H(t)))$, the variance of $\hat{H}(t)$ is immediately available:

$$\operatorname{var}(\hat{H}(t)) = \operatorname{var}(-\log \hat{S}(t)) = \sum_i \frac{d_i}{N_i(N_i - d_i)}. \qquad (10)$$

Cumulative hazard rate curves have a disadvantage in that they highlight the instabilities in the distribution, particularly if the sample is small. However, Fig. 4.4 below provides valuable indices which we will detail in the next chapter.

4.2. Single Sample: Competing Risks

In the preceding demonstration, no distinction was made between the different forms of event studied. Only losses of information—censorings—were taken into account. Let us now suppose that the event that occurs randomly at time T can be of various kinds, identified by J. If, in addition to this, each individual may experience only one occurrence, then competing risks occur. When several causes of death are distinguished, and the individual may die of only one of these causes, then this notion is clearly applicable. The notion of competing risk is present each time the event studied can be distinguished by its form.

The distribution of the couple (T, J) may be considered, where T measures the duration of stay until the moment of occurrence, and J the nature of the event; the conditional density is then equal to

$$h_i(t) = \lim_{\Delta t \to 0} P(T < t + \Delta t, J = i \mid t \leq T), \qquad (11)$$

following

$$h(t) = \sum_i h_i(t); \qquad (12)$$

and, knowing that events occur in t, the probability that they are of type i is

$$h_i(t)/h(t),$$

so that

Study of a Single Event

$$P(J = i) = \int_0^\infty h_i(t) \exp\left[-\int_0^t h(s)ds\right] dt. \qquad (13)$$

An interesting particular case occurs if $h_i(t) = \alpha_i h(t)$, which requires that T and J be independent variables.

Unless risks are considered independent from each other (thus leading to the calculation of survival curves (durations of stay) using Kaplan–Meier estimations for each cause), the problem of interdependence between the different risks arises.

Moreover, when different causes for the occurrence of the same event are envisaged within a sample that is sometimes limited, then it may be feared that the population experiencing the risk may become scarce, and so reliable estimation of the similarities or differences in behaviour pattern may become impossible.

In this case, the cumulative hazard rate estimator, proposed by Aalen (formulas (10) and (11) Chapter 3), is perfectly suitable. If $t_{i1} < t_{i2} < \ldots$ are the instants at which the ith event occurs in the population observed, if $Y(t_{ij})$ is the population at risk just before instant t_{ij}, and if $n(t_{ij})$ individuals experience this event at instant t_{ij}, then

$$\hat{H}_i(t) = \sum_{t_{ij} \leq t} \frac{n(t_{ij})}{Y(t_{ij})} \qquad (14)$$

(the Nelson–Aalen estimator for grouped data), the variance of which is

$$\operatorname{var}(\hat{H}_i(t)) = \sum_{t_{ij} \leq t} \frac{n(t_{ij})}{[Y(t_{ij})]^2}. \qquad (15)$$

Here $Y(t)$ represents a group of individuals at risk just before t and may be modified at each date t without an event necessarily having taken place.

The connection between this estimator and that of Kaplan and Meier is clear. But here the h_is are always specified, even when risks are not independent, so the plot always provides unambiguous information, unlike the Kaplan–Meier survival curves.

The plot $\hat{H}_i(t)$, called the Nelson–Aalen plot, gives the following indications:

1. Their slopes are estimates of the hazard value of each type of event, which may therefore be compared at each date t.

74 *Extending the Scope of Longitudinal Analysis*

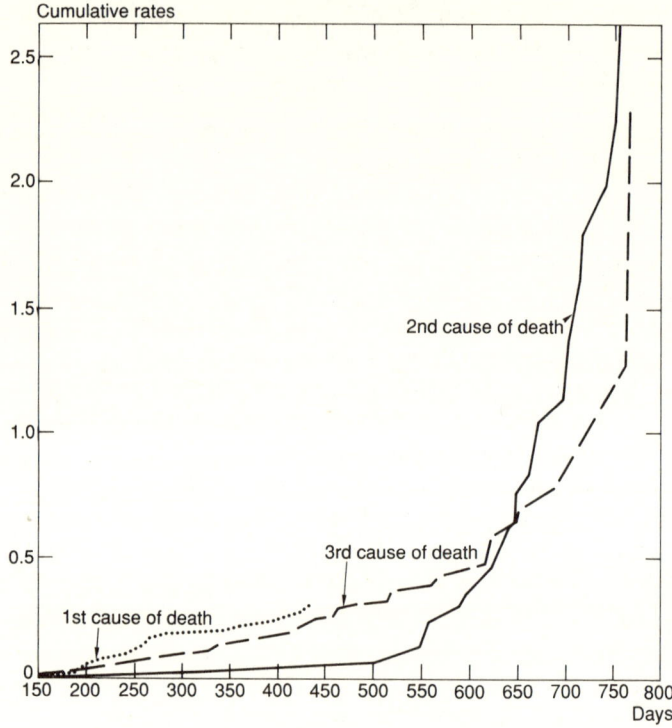

FIG. 4.1. Mortality according to three causes

Cause of death	Percentage of individuals	Survival time (days)
1	23	159 189 191 198 200 207 220 235 245 250 256 261 265 266 280 343 356 383 403 414 428 432
2	40	317 318 399 495 525 536 549 552 554 557 558 571 586 594 596 605 612 621 628 631 636 643 647 648 649 661 663 666 670 695 697 700 705 712 713 738 748 753
3	37	163 179 206 222 228 249 252 282 324 333 341 366 385 407 420 431 441 461 462 482 517 517 524 564 567 586 619 620 621 622 647 651 686 761 763

Note: This example is taken from Aalen (1982). It concerns a sample of mice for which 3 causes of death were distinguished. (It is not useful to name the causes here.)

2. These plots may be used to test parametric distribution of the processes studied.
3. The value of $\hat{H}_i(t)$ at any time is an estimate of the expected number of occurrences of i type that would have occurred if a single individual were constantly at risk. It is not a common quantity to be considered.

Examples
1. The following example shows the cumulative mortality intensity plots showing three causes of death in a sample of 95 individuals.

Figure 4.1 shows cumulative hazard curves. It shows that the 1st and 3rd causes of death occur with constant and comparable probability at the beginning of the period. By the time the 1st cause has disappeared, the 2nd cause has become more significant. The 3rd cause thereupon increases in importance. (The hazard curve suddenly slopes upward.) However, after 600 days the 2nd cause predominates.
2. With this model, observation $Y(t)$ may be dealt with in a multiplicative way. So it can also be estimated using other methods than simply counting the individuals at risk. For example, in the case of an epidemiological study, Aalen suggests modelling the infectiousness in introducing the number of infectives $c(t)$ and the number of susceptibles $n(t)$ to contract the illness. In this case the multiplicative hazard model may be written

$$\Lambda(t) = h(t) c(t) n(t),$$

where $h(t)$ is a measure of the individual's infectiousness and

$$\hat{H}(t) = \sum_{t_i \leq t} \frac{1}{c(t_i) n(t_i)}.$$

4.3. Multiple Samples: Comparative Tests

If several samples are available, one is often led to determine whether the same distribution governs the occurrences observed in the different samples. The test used comes from rank testing techniques. This is because we are in the presence of a comparison of k distributions. The zero hypothesis corresponds to the fact that these k distributions are identical, and of unknown density. Log-rank tests are therefore used; these use the serial number of the observations only when the latter[1] are in rising order.

Let there be a sample which may be divided up into k distinct sub-populations, and let $t_1 < t_2 < \ldots < t_n$ be the dates when the event studied occurs. Let us also suppose that d_j occurrences are recorded in t_j, while N_j individuals are at risk just before $t_j (j = 1, \ldots, r)$. Moreover, the number of individuals and occurrences corresponding to the different sub-populations will be noted d_{ij} and N_{ij}, with $(i = 1, \ldots, k)$. The distribution of $d_{ij}, \ldots, d_{kj}$ is then the product of binominals:

$$\prod_{i=1}^{k} \binom{N_{ij}}{d_{ij}} h_j^{d_j} (1 - h_j)^{N_j - d_j}, \tag{16}$$

where h_j is the hazard rate common to k distributions. Consequently, the distribution of $d_{ij}, \ldots, d_{kj}$ is hypergeometric, given d_j:

$$\frac{\prod_{i=1}^{k} \binom{N_{ij}}{d_{ij}}}{\binom{N_j}{d_i}}. \tag{17}$$

The mean of the d_{ij} may be inferred:

$$w_{ij} = N_{ij} \frac{d_j}{N_j}; \tag{18}$$

the variance is

$$(V_j)_{ii} = \frac{N_{ij}(N_j - N_{ij}) d_j (N_j - d_j)}{N_j^2 (N_j - 1)}; \tag{19}$$

and the covariance of d_{ij} and d_{lj} are

$$(V_j)_{il} = \frac{-N_{ij} N_{lj} d_j (N_j - d_j)}{N_j^2 (N_j - 1)}; \tag{20}$$

The transpose of the test statistic v_j is $(d_{1j} - w_{1j}, \ldots, d_{kj} - w_{kj})$, a vector with zero mean and with variance–covariance matrix equal to V_j.

Vector $v = \Sigma_1^k v_j$, which is composed of the occurrence observed in each sub-population minus the corresponding number of expected occurrences, constitutes the rank statistic.

---

[1] We refer here to survival times.

Study of a Single Event

Distributions are equal when, asymptotically, $v' V^{-1} v$ is a χ^2_{k-1}. The statistic of χ^2_{k-1} is formed by using only $k-1$ elements, since, given $k-1$ elements, the kth is perfectly defined. In fact, the sum of k elements in v is 0.

Examples

Before examining examples of analyses from the 'Triple Biographie' survey, let us take a 'microscopic' application as an exercise, as we did earlier.

Take the population of two buildings of identical status located in two different parts of a town. We shall test the possible differences to be found in out-migrations of the inhabitants.

Thus at instant t_6, over the two sets, the out-migration hazard rate is $q_6 = 5/21 = 0.02381$. If this hazard rate were applied to each building separately, 1.667 out-migrants from the first and 3.333 from the second building would be observed (see Table 4.1). Since 0 and 5 respectively are observed, the test statistic

$$v_6 = \begin{pmatrix} 0 - 1.667 \\ 5 - 3.333 \end{pmatrix} = \begin{pmatrix} -1.667 \\ 1.667 \end{pmatrix}$$

and

$$v = \sum_{j=1}^{10} v_j = \begin{pmatrix} -1.886 \\ 1.886 \end{pmatrix},$$

TABLE 4.1. Out-migration of two fictitious populations

Duration	First population			Second population				
t_j	d_{1j}	N_{1j}	w_{1j}	d_{2j}	N_{2j}	w_{2j}	d_j	N_j
t_1	4	19	2.375	1	21	2.625	5	40
t_2	1	15	0.429		20	0.571	1	35
t_3		14	0.824	2	20	1.176	2	34
t_4	3	14	1.931	1	15	2.069	4	29
t_5	2	10	0.833		14	1.167	2	24
t_6		7	1.667	5	14	3.333	5	21
t_7		6	1.600	4	9	2.400	4	15
t_8	1	6	2.727	4	5	2.273	5	11
t_9		1	0.500	1	1	0.500	1	2
t_{10}	1	1	1.000					

the variance–covariance matrix of which is

$$V = \begin{pmatrix} 5.828 & -5.828 \\ -5.828 & 5.828 \end{pmatrix}.$$

The statistic,[2] which is calculated and which is compared to a χ_1^2, is of value $(1.886)^2(5.828)^{-1} = 0.610$, which is not significant. It may therefore be concluded that no difference appears between out-migration of the occupants of one building in the study as compared with the other.

Similar non-parametric analyses may be made using two packages: 'Life Tables and Survival Functions 1 L' from BMDP (PL1), and LIFETEST from SAS. These two programs may be used to estimate durations of stay, hazard rates, and density distribution, and to make comparisons between distinct sub-populations in the sample. Moreover, they produce plôts.

The example presented next concerns the study of individuals' professional careers, and in particular career mobility. Here, M. A. Cambois has used LIFETEST from SAS. Durations of stay are analysed by taking account of the occupational status categories defined in 1975 by the French INSEE nomenclature: farmers; farm labourers;

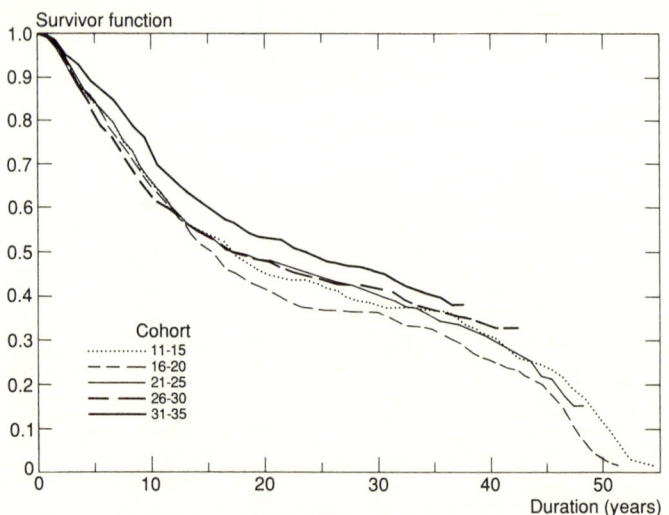

FIG. 4.2. Survivor function in the first job in years: males

[2] Care should be taken in forming the test statistic, and only $k-1$ elements, i.e. only one in this case, should be taken into account.

self-employed; managerial staff; executive staff; other staff employees; manual workers; service personnel; artists, clergy, the army, and the police.

The study of the durations of stay for males in the first occupational group (Fig. 4.2–4.4) shows greater stability for the most recent cohorts. Three types of curve are shown here for the same analysis: that

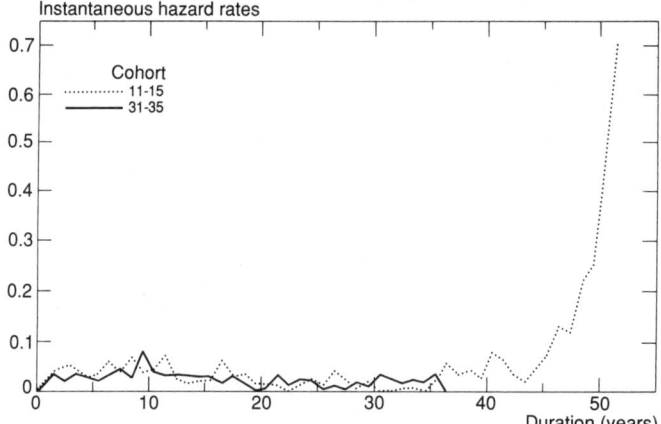

FIG. 4.3. Exit of the first job: males

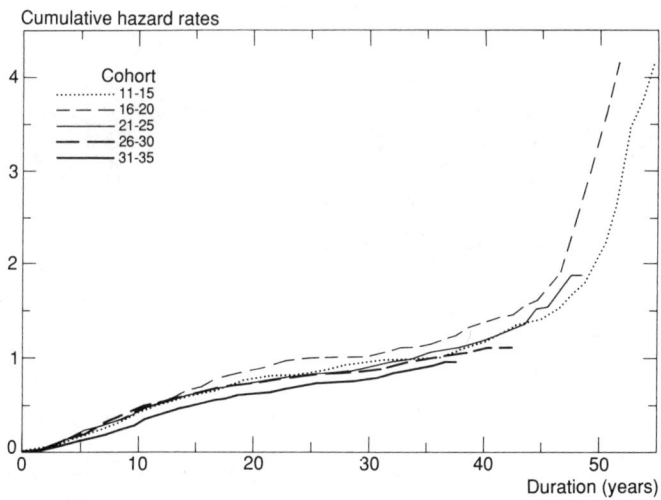

FIG. 4.4. Cumulative hazard rates, exit of the first job: males

of durations of stay, that of hazard rates (only for two cohorts), and finally, the cumulative hazard rate plot. It will be noticed that cumulative hazard rates may exceed 1 (See Section 2.1.3).

Figure 4.5 shows the distribution for men and women in the first occupational category. Both men and women leave a first job as a farmworker as quickly as possible. On the other hand, if the first job is as a worker, the behaviour pattern is very different between sexes.

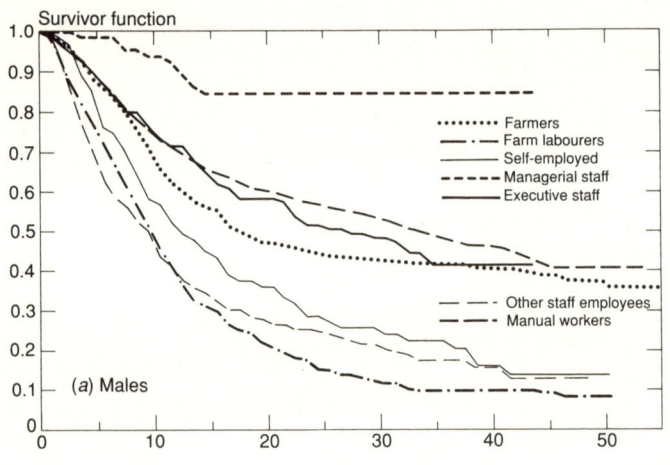

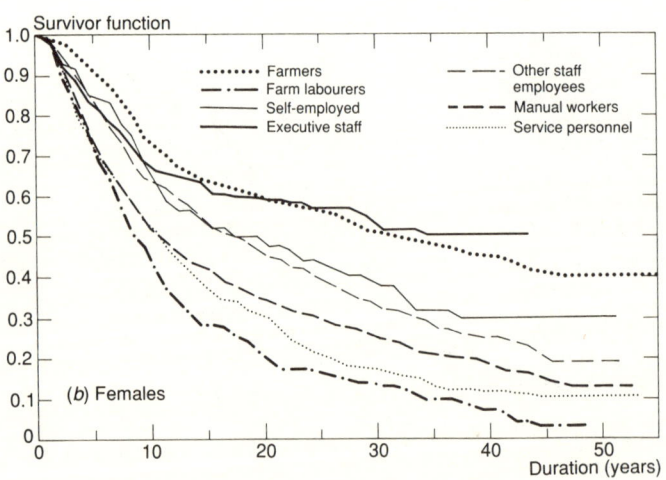

FIG. 4.5. Duration of stay in the initial group

Women who have started as workers leave this status almost as quickly as farmworkers or service personnel do, whereas, after the executive or liberal professions, it is the category in which men remain the longest.

4.4. Conclusion

In this chapter we have examined the different analyses that might be made when a single process is studied. The non-parametric methods used here provide estimators that are very useful as it is always possible to use their variance. It is thus possible to judge the goodness of fit of the estimators, and so comparisons may be undertaken.

This happens when the process studied is expressed according to different categories, such as causes of death, or when a comparison between different sub-populations is made.

This chapter has also enabled us to highlight an essential difference between this analysis and a classical one. It is not necessary, here, to pose the awkward hypothesis of studying the process without censoring.

These methods applied to several processes open up new avenues. We shall now, therefore, examine the analysis of interaction between demographic phenomena.

5
Reciprocal Study of Interactions between Two Events

We shall now examine the bivariate case, which makes it possible to study the interactions between two events—marriage and departure from the farming community, for example. We shall present the concepts underlying the analysis before formalizing it.

5.1. Conception of the Analysis

5.1.1. Two competing phenomena

In the univariate case, the problems posed by competing risks arise when only one occurrence is observed out of all the possible occurrences. However, study of behaviour patterns leads to the consideration of the more complex aspects of individual life-courses. As was seen in the introduction, individuals are not confronted with single choices: rather, their decisions are the result of arbitrage processes, whether objective or not, between systems in which they are implicated, and among which coherence must be maintained. Thus, if a demographic phenomenon is studied independently of its context, the analysis will be oversimplifying. In the same way, it is not enough to study the influence of a series of factors on a phenomenon. It is necessary to take account of the distribution of several competing phenomena in the course of an individual's life.

We shall first of all examine the case of two events, and study the reciprocal influence of one upon the other. It would, indeed, be simplistic to sketch out the stages of a life-cycle according to a hierarchy leading to the final state (death). Our approach avoids this reduction and confronts the events of various fields. The life-cycle is formed of stages marked by events of varying kinds, which do not follow a pre-established order. Thus, accession to the adult state is usually

Study of Interactions between Two Events

marked, among other things, by departure from the parents' home, marriage, a first job, the establishment of firm political views, and so on—in other words, events in the family, residential, professional, or socio-political fields. An individual does not necessarily go through all these stages, and there is no pre-established order. Therefore, for each person, different areas may be observed in which individual life-cycles unfold, and the final coherence of each cycle depends on various individual adjustments. The object of the analysis is to explore these adjustments, and to determine the dependences created between the different systems.

5.1.2. Dynamic differentiation

Rather than speak of correlations between phenomena, we prefer to refer to their interactions, since we are going to envisage various forms of dependence. We therefore intend to analyse the behaviour patterns that arise from confrontation between the different on-the-spot demands. Consequently, we shall not here study the differences between sub-populations designated by one characteristic or set of characteristics, but rather shall identify how the members of the population in an initial homogeneous cohort differ in the course of the life-cycle according to the individual manner in which they adapt, whether by obligation or free will.

The study of preferential decisions, of obligatory stages, and of favourable paths becomes the basis for an approach to behaviour patterns which can be understood only through the dynamics of the processes.

In our case, having at our disposal a population group assumed to be homogeneous at the beginning of the period, our tool will enable us to explore the way in which this group evolves, and the way in which heterogeneity appears among its members. It may even enable us to see that heterogeneity existed right from the start. The differences that interest us are those that appear during the life-cycle, because of interactions between the different areas in which the individuals are implicated.

5.1.3. The different types of dependence

Let T_1 and T_2 be the times at which two events occur. The study consists in analysing the interactions. They are measured by the effect observed on the distribution of the first event when the second occurs and, in return, the changes wrought upon the distribution of the second by the appearance of the first. The process is shown schematically in Fig. 5.1.

The equality test between the 'before' and 'after' instantaneous failure rates gives an indication on the stochastic dependence of the first event on the second, without presupposing that the inverse is true.

Thus, by dissociating the dependences, the meaning of the influences may be determined in probabilistic terms and not in deterministic causality terms.

In the applications that will be described in this chapter, never once did we arrive at a finding of total independence between the two phenomena studied, which would correspond to verifying the two equalities. The danger of handling demographic phenomena separately, on the hypothesis that they are independent of each other, may thus clearly be seen.

On the other hand, if only one of the two equalities is verified, unilateral dependence (or local dependence) is clearly shown. Thus, French men born between 1911 and 1935 and having started in a farming job get married far more frequently once they have left that job, whereas their change of employment does not depend on their marital status. The opposite is true for women, whom marriage tends to fix in farming jobs, whereas leaving the farm does not modify their chances of getting married (Courgeau and Lelièvre, 1986). Similarly,

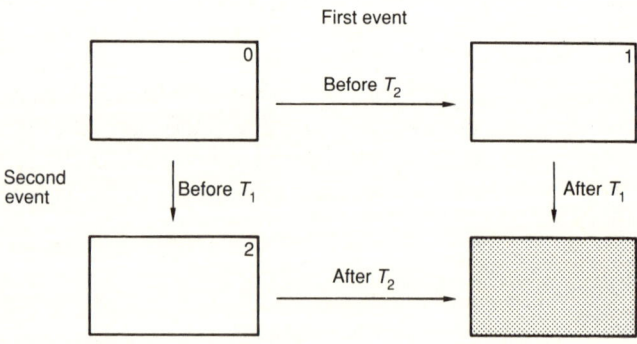

FIG. 5.1

the purchase of a first house does not depend on the birth of the last child (of completed families) for worker couples; on the other hand, home ownership does tend to make the birth of a last child more probable (Courgeau and Lelièvre, 1988b).

These results clearly show the relative importance of the two areas of which competition is tested for the group studied. These dependences, which seem parodic at the stage of non-parametric analysis, will be the object of a second, semi-parametric or parametric approach, which will enable their exact nature to be shown. Moreover, the nature of the sub-groups that have stood out will then be identified by examining the individual characteristics of their members.

Finally, when none of the equalities tested is verified, the two phenomena are said to be in reciprocal dependence. This is the case for marriage and departure from the parents' home, for generations born between 1926 and 1935 (Courgeau et al., 1986). Similarly, while the probability of migration to metropolitan areas diminishes after a birth, second and further births are reduced after the same kind of migration (Courgeau, 1987c).

These analyses have enabled us to work on small samples or on highly stratified data. We were able to test previous equalities, not only over the entire period studied, but also at each period or at each age.

Dependences may be observed only at certain ages, or over a given period after the modifying event. The influence may also reverse itself. Thus, when fertility and female activity is studied, it is observed that women without a job at marriage and when the first child is born have fertility behaviour patterns that differ markedly depending on age: although before age 30 those who re-enter the labour force are as fertile as those without a job, if they return to working life after they are 30, further births become rare (Lelièvre, 1987a).

Finally, other more complex levels of interpretation may be revealed. When interactions between fertility and migration to metropolitan areas were studied, a spectacular reduction of fertility, of rank 2 or more, was observed in female migrants. The question to be asked is, Is this an adaptation behaviour, or a selection, in so far as the sample of future female migrants already has a fertility behaviour pattern that differs from that of other women in the region of origin? With the use of event history data, it is possible to test these differences between future female migrants and sedentary women in the population of the original area of residence.

In fact, it was verified that future migrants to metropolitan areas already have a low fertility rate for second and third births compared with sedentary women in zones of non-metropolitan areas.

A *priori* dependence of fertility on future migration is also displayed, directly connected to this selection within the initial population.

Reciprocally, a favourable effect on second or third births is produced by migration to non-metropolitan areas. An identical investigation had enabled us to show genuine adaptation of fertility behaviour patterns in those women who migrate from metropolitan areas. Their earlier fertility behaviour pattern in no way differs from that of city-dwellers who do not leave metropolitan areas.

5.1.4. Unilateral dependence and causality

We have revealed four types of dependence or independence between phenomena. It is tempting to say that these dependences are the cause of the phenomena. But, over and above a causality reasoning, even in a probabilistic and non-deterministic sense, we have analysed them in terms of interactions and possible reciprocity.

We shall show here how this analysis differs from a causal analysis. To situate this approach in comparison with current causal explanation practice, we shall examine the probabilistic theories. The definition given by theoreticians of probabilistic causality is as follows:

c and e being specific observed events, c is the cause of e if, and only if,

1. c does not occur after e;
2. $P(e \mid c) > P(e)$;
3. there exists no event s such that s is earlier than or simultaneous to c and s screens off c from e; s screens off c if $P(e \mid s,c) = P(e \mid s) \neq P(e \mid c)$; intuitively, s is then the true cause of e, of which c is only an indice. (Good, Salmon, and Suppes; quoted in Swain, 1987)

The study of the interactions that we recommend is characterized by the identification of *unilateral dependences*. The relation between two phenomena is thus dissected using a systematic reciprocal approach. Our method therefore never excludes the possible reciprocity of influence, and is, at first, in contradiction with the anti-symmetry assumption that underlies the probabilistic definition: if c is the cause of e, then e clearly cannot be the cause of c. Here our reasoning is definitely of dependence and not causality.

Secondly, we do not keep axiom 1, which is common indeed to all definitions of causal relationships, and which states the temporal anteriority of the cause. The idea of reciprocity invalidates the form of temporality defined in the axiom. Moreover, as we have seen, we succeed in determining a *priori* behaviour patterns, such as selection phenomena which favour a particular type of migration. This analysis is capable of identifying complicated forms of temporality.

We shall therefore consider our model as being outside the framework of causal explanation, and shall consider it rather as an analysis of phenomena and their interactions. This analysis genuinely develops the classical demographic analysis of phenomena in their 'pure state', by introducing the analysis of phenomena in interaction.

For the following discussion we shall use a bivariate study. It may be, for example, the analysis of nuptiality among farmworkers (Courgeau and Lelièvre, 1986) with a study of the interactions between marriage and departure from the farming community; or, again, an analysis of the interactions between fertility and female activity (Lelièvre, 1987a), between fertility and migrations to or from metropolitan areas (Courgeau, 1987c), or between home ownership and the last birth (Courgeau and Lelièvre, 1988b). On each occasion, we shall study the interactions between two events, the arrival of which result from an arbitrage, or an objective or subjective choice of priorities; and we shall then try to identify the various types of dependence between the two phenomena.

5.2. The Bivariate Case

Conditional distributions are formalized as follows. Let T_1 and T_2 be the random times at which two events occur. The hazard rates (Fig. 5.2) are therefore defined by

$$h_{01}(t) = \lim_{\Delta t \to 0} P(T_1 < t + \Delta t \mid T_1 \geq t, T_2 \geq t), \qquad (1)$$

which is the instantaneous failure rate for the first event if the second has not yet occurred, and by

$$h_{21}(t \mid u) = \lim_{\Delta t \to 0} \frac{1}{\Delta t} P(T_1 < t + \Delta t \mid T_2 = u, T_1 \geq t), \qquad (2)$$

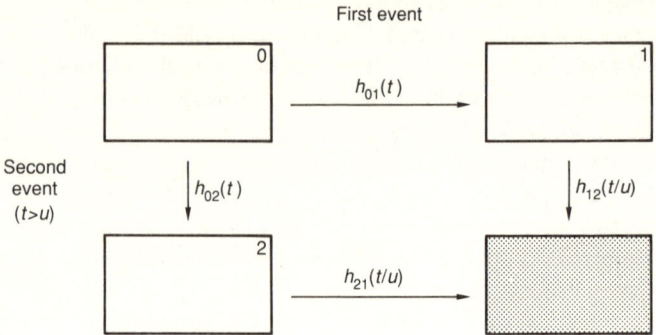

FIG. 5.2. Diagram of states in bivariate case $(t > u)$

which is instantaneous failure rate for the first event if the second occurs before date u.[1]

The equality test between h_{01} and h_{21}, and that between h_{02} and h_{12}, then gives an indication of the stochastic dependence of the random variable T_1 in relation to T_2:

- reciprocal dependence of the two events if $h_{01}(t) \neq h_{21}(t \mid u)$ and $h_{02}(t) \neq h_{12}(t \mid u)$;
- unilateral local dependence if $h_{01}(t) \neq h_{21}(t \mid u)$ and $h_{02}(t) = h_{12}(t \mid u)$, thus showing at the same time whether it has a negative or a favourable effect on the occurrence of T_1, depending on whether $h_{01}(t) > h_{21}(t \mid u)$ or $h_{01}(t) < h_{21}(t \mid u)$; and, finally,
- total independence of the two events if $h_{01}(t) = h_{21}(t \mid u)$ and $h_{02}(t) = h_{12}(t \mid u)$.

Following this, in continuous time the density of the couple of random variables (T_1, T_2) may be established, given that the survivor function of T_i is

$$S_i(t) = \exp\left(-\int_0^t h_{0i}(u)du\right) \qquad (3)$$

and its probability density:

$$f_i(t) = h_{0i}(t) S_i(t). \qquad (4)$$

---

[1] Symmetrically, $h_{02}(t)$ is obtained, and

$$h_{12}(t \mid u) = \lim_{\Delta t \to 0} \frac{1}{\Delta t} P(T_2 < t + \Delta t \mid T_1 = u, T_2 \geq t).$$

If $f_2(t_2 \mid t_1)$ is the conditional density of T_2, given that $T_1 = t_1$, then $f_1(t_1)$ is the marginal density of T_1. We will still have $f(t_1, t_2) = f(t_2 \mid t_1) f(t_1)$. From this, it is therefore possible in our model to write the probability of remaining in I_1 until date t_2:[2]

$$f_2(t_2 \mid t_1) = h_{12}(t_2 \mid t_1) \exp\left(-\int_{t_1}^{t_2} h_{12}(u \mid t_1) du\right), \quad (5)$$

and the probability of remaining in I until date t_1:

$$f_1(t_1) = h_{01}(t_1) \exp\left(-\int_0^{t_1} (h_{01}(u) + h_{02}(u)) du\right). \quad (6)$$

If $t_1 \leq t_2$, before t_1 one can only be in I and there are two ways of leaving it; so

$$f(t_1, t_2) = h_{01}(t_1) h_{12}(t_2 \mid t_1) \exp\left[-\int_0^{t_1} (h_{01}(u) + h_{02}(u)) du \right.$$
$$\left. - \int_{t_1}^{t_2} h_{12}(u \mid t_1) du \right], \quad (7)$$

with a similar expression for $t_2 \leq t_1$.

Now, T_1 and T_2 are independent if $f(t_1, t_2) = f_1(t_1) f_2(t_2)$. A necessary and sufficient condition may be expressed in the form

$$h_{21}(t \mid u) = h_{01}(t) \text{ and } h_{12}(t \mid u) = h_{02}(t).$$

Should the second equality be verified, the joint density function (7) then becomes

$$f(t_1, t_2) = h_{01}(t_1) \exp\left(-\int_0^{t_1} (h_{01}(u) + h_{02}(u)) du\right)$$
$$\times h_{02}(t_2) \exp\left(-\int_{t_1}^{t_2} h_{02}(u) du\right). \quad (8)$$

The imbalance of the formula is because the first event occurs before the second.

---

[2] I_1 being the state in which individuals having experienced the first event are found, and I being the initial state.

5.2.1. Actuarial estimation

To estimate these hazard rates, we assume them to be constant throughout a year. This non-parametric version in discrete time makes estimation possible when no purely non-parametric method is available or satisfactory (Cox and Oakes, 1984). It is also assumed that during these intervals the events considered (for instance, marriages and departures from farming communities) are uniformly distributed.

Thus, let $N_i(t)$, $i = 0, 1, 2$, be the population in state i at the start of year t, and $n_{ij}(t)$ be the number of events of type j occurring in the population of state i during year t. The most simple estimators are given by

$$\hat{h}_{0,1}(t) = \frac{n_{0,1}(t)}{N_0(t) - \frac{1}{2}[n_{0,1}(t) + n_{0,2}(t)]} \tag{9}$$

$$\hat{h}_{2,1}(t \mid u) = \frac{n_{2,1}(t)}{N_2(t) - \frac{1}{2}[n_{2,1}(t) - n_{0,2}(t)]} \tag{10}$$

$$\hat{h}_{0,2}(t) = \frac{n_{0,2}(t)}{N_0(t) - \frac{1}{2}[n_{0,2}(t) + n_{0,1}(t)]} \tag{11}$$

$$\hat{h}_{1,2}(t \mid u) = \frac{n_{1,2}(t)}{N_1(t) - \frac{1}{2}[n_{1,2}(t) - n_{0,1}(t)]} \; . \tag{12}$$

In our example, to eliminate simultaneities we have made the following hypothesis: individuals married the year they left the farming sector are studied with the community of married farmworkers for the calculation of exit from agriculture hazard rates, and with the population of bachelors having left farmwork for that of nuptiality hazard rates.

There are other ways of taking account of simultaneities, which will be detailed later.

5.2.2. Possible tests

One of the problems of this analysis is to test the hypotheses that determine the particular dependences between phenomena. Often, there are reduced numbers of individuals, so that all the possible tests cannot be used.

The most simple test is a generalized rank test proposed by Peto and Pike (1973) and used by Aalen et al. (1980). The other tests are based on the normalized differences between observed and estimated values and the calculation of the variance–covariance matrix of the two distributions.

With this test, however, only a general comparison between the trends represented by two series of hazard rates h_{0i} and h_{ji} is possible. It is a non-parametric test with two components. Here, the occurrence of one of the events, depending on whether the second has occurred or not, is compared.

Let

$t^1_{\cdot}$ be the dates of transitions from 0 to i
$t^2_{\cdot}$ be the dates of transitions from j to i
$Y_k(t)$ be the population at risk in state k at instant t.

Savage's statistic S is then calculated thus:

$$S = \sum_k \frac{Y_j(t^1_k)}{Y_0(t^1_k) + Y_j(t^1_k)} - \sum_k \frac{Y_0(t^2_k)}{Y_0(t^2_k) + Y_j(t^2_k)}, \qquad (13)$$

a non-biased estimator of its variance, V:

$$V = \sum_k \frac{Y_0(t^1_k) Y_j(t^1_k)}{[Y_0(t^1_k) + Y_j(t^1_k)]^2} + \sum_k \frac{Y_0(t^2_k) Y_j(t^2_k)}{[Y_0(t^2_k) + Y_j(t^2_k)]^2}. \qquad (14)$$

Under the equality hypothesis between the two series, statistic $SV^{-1/2}$ is then asymptotically normally distributed according to $N(0, 1)$.

However, S can measure differences on intervals only where Y_0 and Y_j are simultaneously different from zero.

5.2.3. Tests chosen and the convergence of estimations

To test the equality hypotheses of hazard rates, we prefer to use test statistics based on the differences between the estimators of the two hazard rates. In fact, these estimators all come from the asymptotic theory of occurrence on risk rates, which J. Hoem (1976) reviewed for the field of demography.

Whatever the distribution model of the phenomenon, the actuarial estimator, given the hypothesis that the hazard rate remains constant throughout the interval, will have the following form:

$$(h - \hat{h})/(- d^2 \log (h)/dh^2), \tag{15}$$

which has a normal asymptotic behaviour and so enables confidence intervals to be computed.

On these statistics, Schou and Vaeth (1980) have made simulations to determine the minimum size of total numbers which will allow them to keep their asymptotic convergence. To this extent, the choice of these estimators seemed to us to be the most consistent.

Suppose that we want to know if the hazard rate corresponding to the first event remains unchanged when the second occurs. We will want to test, for date t, whether the following equality is verified:

$$\hat{h}_{01}(t) = \hat{h}_{21}(t) .$$

Let $Y_0(t)$ and $Y_2(t)$ denote the populations at risk of the first event, depending on whether they have not experienced the second or, on the contrary, have already experienced it.

When the population observed tends towards infinity, $Y_0(t) \to \infty$, $Y_2(t) \to \infty$, then

$$D(t) = (\hat{h}_{01}(t) - \hat{h}_{21}(t)) \bigg/ \left(\frac{\hat{h}_{01}(t)}{Y_0(t)} + \frac{\hat{h}_{21}(t)}{Y_2(t)} \right)^{1/2}, \tag{16}$$

where the denominator is a consistent estimator of the asymptotic standard deviation of the numerator, taking account of the respective size of the populations at risk, and is asymptotically normal, $N(0, 1)$, if the preceding equality is verified. It is thus possible to verify this equality.

Statistic $D(t)$ may be cumulated over a series of periods, and in this case the test statistic

$$D = \sum_t D(t)/\sqrt{m}, \tag{17}$$

where m is the number of periods taken into account, is asymptotically normal, $N(0, 1)$, if the two sub-populations have the same behaviour pattern throughout the period.

The second estimator uses the result obtained by Schou and Vaeth (1980), who indicate that variables $(h_{01}(t))^{1/3}$ and $(h_{21}(t))^{1/3}$ tend towards their normal distribution faster than the preceding ones. Let $n_{01}(t)$ and $n_{21}(t)$ denote the numbers of individuals having experienced the first occurrence depending on whether they have or have not yet experienced the second.

When the population observed tends towards infinity, then

$$SV(t) = [\hat{h}_{01}(t)^{1/3} - \hat{h}_{21}(t)^{1/3}] \bigg/ \left(\frac{\hat{h}_{01}(t)^{2/3}}{9n_{01}(t)} + \frac{\hat{h}_{21}(t)^{2/3}}{9n_{21}(t)} \right)^{1/2}, \qquad (18)$$

where the denominator is a consistent estimator of the asymptotic standard deviation of the numerator, and is asymptotically normal, $N(0, 1)$, if the equality of the hazard rates is verified.

This statistic may be cumulated over a series of periods as was done for $D(t)$,

$$SV = \sum_t SV(t)/\sqrt{m} \qquad (19)$$

being the number of periods.

From the simulation study made by Schou and Vaeth on the reliability of the estimators of the maximum likelihood in the case of longitudinal data, it is clear that the limitations in the application of the normality hypothesis remain valid as long as $N\hat{h}_k \geq 10$. This result is important when analysing small samples.

The very swift convergence of these estimations towards normal asymptotic distribution shown by Schou and Vaeth's simulation is not, however, verifiable if the estimated hazard rates are nil. In this case, which is relatively frequent when analysing human phenomena, it is impossible to test differences and establish confidence intervals.

5.3. Practical Analysis

The analysis of the interactions between two events covers many different applications. We shall therefore describe the procedure to be followed in the different situations.

The first imperative is to again take account of events quoted in chronological order on an identical scale; the waiting times between each of the occurrences must therefore be measured according to a common origin. Individuals are taken at the start of their professional lives, on their departure from their parents' homes, at their marriage, and so on.

The second imperative to respect is to choose events that do not belong to a series of movements into graded states: it is not possible to study the interactions between the first and the second migration since these necessarily occur in a given order, the first preceding the second by definition. Similarly, taking marriage and divorce, one is obliged to conclude that the main cause of divorce is marriage.

Once these two principles have been respected and two events chosen, many situations remain:

- Some individuals never experience one of the events. Thus, part of the population will always remain single, so it is necessary to take account of right-censoring.
- Some events are reversible, which means that the states to which they lead recur. In this case reversals must be taken into account. It is after all possible to leave farming and return to it, or to have a job and then be unemployed.
- In studying the interactions between different areas of an individual's life, it is desirable to confront the family life-cycle, such as departure from the parents' home, cohabitation, marriage, or successive births, with a single event, such as the first job. There is then a chronological series of events to be analysed, which interact with a single event or with another chronological series.

Naturally, many situations arise which we will not detail here.

The second case we have indicated—that of reversible demographic events—is an interesting one to describe. Under this heading, to-ing and fro-ing between a job and inactivity, or migrations to and from metropolitan areas areas are considered, for example. These transitions correspond in this case to repeated movements into recurrent states. Two analyses may therefore be made: analysis of the interactions between 'to-ing' and the second event, and analysis of the interactions between 'fro-ing' and this second event. Then the instantaneous failure rates of the second event are compared depending on the individual's initial situation.

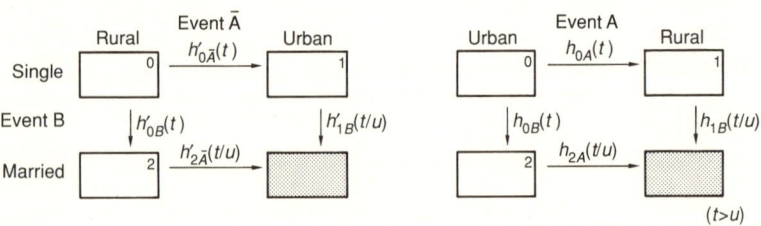

FIG. 5.3. Bivariate case with reversible occurrence

Figure 5.3 depicts the interaction between metropolitan/non-metropolitan migrations and nuptiality. First, interactions between migrations towards urban centres and marriage are considered: individuals follow the direction of the arrows corresponding to hazard rates $h'\cdot_B(\cdot)$ and $h'\cdot_A(\cdot)$

Then the interactions between migrations from metropolitan to non-metropolitan areas and marriages are analysed: individuals follow the direction of arrows $h\cdot_B(\cdot)$ and $h\cdot_A(\cdot)$.

Finally, it will be possible to compare the nuptiality hazard rates of city-dwellers $h_{0B}(t)$ and $h'_{1B}(t \mid u)$ two by two, depending on whether they are originally from the metropolitan areas or in-migrated there, having originated in non-metropolitan areas. Similarly, nuptiality hazard rates $h'_{0B}(t)$ and $h_{1B}(t \mid u)$ for those who live outside large cities may be compared, depending on whether or not they originated in the metropolitan area.

Such applications cannot be envisaged without an efficient software program to carry them out. A package like this was therefore designed at the Institut National d'Etudes Démographiques (INED) by E. Lelièvre, making it possible to test the different working hypotheses presented here. The Root program was used for all the applications described and is presented in the Appendix.

5.3.1. Simultaneities and time intervals

In studying interactions, events are found to be competing at a particular moment of the life-cycle. However, it may be difficult to establish a scale of times that makes an analysis possible. Depending on the time-scale chosen, the researcher may observe cases of simultaneity between two events, such as marriage and migration at the same age, or leaving a job and giving birth at the same date. Now, while simultaneous events are not a problem in mathematics—in continuous time it is always possible to detect the anteriority of one occurrence over the other—in social sciences the problem is different, since the events observed are but the realization of previous decisions of which the date is unknown.

Thus, even if data are collected meticulously (to the day, hour, and minute), and it is possible to dissociate for every case the occurrence of two events, how valid is this distinction? When the statistician eliminates the possibility of simultaneous occurrences, what can the demographer or the sociologist deduce from two events one month apart, and experienced as simultaneous?

The distinction between events recorded as simultaneous may be made only if the maturation and the circumstances of the event, and how it was decided upon, are known. Without this information, the demographer may choose to act as a statistician or decide to take account of these uncertain situations. This is why we have introduced the notion of 'fuzzy time' (Courgeau and Lelièvre, 1988a), which characterizes the time elapsed between the maturation (or possible decision) and realization (or observed occurrence) of the event. This time lapse is different for each individual, and is unknown to the demographer.

The two occurrences are said to be simultaneous when they occur in a chosen time interval. (Two events may be considered as simultaneous if they occur in the same quarter, or the same year.) Because of the 'fuzzy time' notion, the researcher does not allow himself to choose, and does not decide whether an event in the chosen time-frame is anterior to another.

This controversial method of considering simultaneities on their own does not facilitate the statistical approach to the problem, but rather calls directly for pluri-disciplinary collaboration. This is because, in this case, the set of simultaneous movements must be characterized in a similar way to that of joint decisions. But only psychologists and sociologists working on maturation of projects and the logic of directions are capable of integrating interpretation.

The work of E. Klijzing, J. Siegers, N. Keilman, and L. Groot (1988) must be mentioned in this connection. In their study on departures from professional activity and births, these researchers experimented to define the most suitable time-scale for decision-making in cases of simultaneity between departure from work and births.

Their method consisted in taking intervals of various sizes (from one to nine months); and a birth and a departure from work were considered as simultaneous if they occurred within the same interval. Therefore, it was considered that, the longer the interval, the more simultaneities there would be. The authors then proceeded to test equality, $h_{01}(t) = h_{21}(t \mid u)$ in the case of having a child, whether as a working mother or not (Table 5.1), excluding simultaneities. It may be noted that hazard rates remain fairly stable, but above all, the test statistic remains virtually unchanged. Within an interval, kept in this case to nine months maximum, the estimation of the influence of one

TABLE 5.1. Tests of 'fuzzy time'

2 years' fertility hazards	Tie widths (months)			
	1	3	6	9
h_{01} : active women	5.5	5.3	5.1	5.0
h_{21} : women inactive again	23.2	23.7	22.1	24.1
Test statistic $N(0, 1)$ measuring the significance of difference between h_{01} and h_{21}	−3.311	−3.401	−3.264	−3.495

Source: Klijzing et al. (1988)

event upon the other remains the same whether the dates of the two events are distinguished or not.

In the Root program, we have allowed for taking account of simultaneities according to eight options. These options allow the researcher to substract, or incorporate into the analysis, cases where the two events occur in the same time interval, or, again, to calculate the corresponding hazard rates separately.

The hypothesis of being at risk during a half-interval, which concerns those individuals who experience an event during this interval, is that retained in all the applications. This hypothesis assumes that events are distributed uniformly over the interval.

5.3.2. Plots

As was pointed out in the previous chapter, the plots used by the researcher to illustrate the analysis may be of several types. We shall therefore discuss here the various possibilities, evaluating their advantages and disadvantages. Several statistics are available: the hazard rates, the cumulative hazard rates, and the survivor function. These are represented in Figs. 4.2–4.4 above. In the papers we have published, we have chosen mainly to present cumulative hazard rates. These are questionable, however, in so far as the apparent 'level' of the phenomenon connects to no notion of interest. In fact, this confusion comes from the fact that the accumulation of hazard rates may exceed the limit value 1, since we are not talking about a sum of probabilities over a stable population, but rather of a sum of risks.

(A risk may itself be superior to 1; cf. Section 2.1.3 above.) In fact, only the differences (differences in slope) between the curves indicate divergences of behaviour patterns, and it is these that must be considered. Moreover, a cumulative hazard rate plot gives clear indications of the parametric form that should be chosen to modelize the processes studied.

The survivor function (see Section 4.3) is less difficult for the reader to use. The curve decreases from the unit to minimum values corresponding to the proportion of those who have still not experienced the event at the end of the observation. However, these curves are obtained by calculations using the hazard rates, and it is likely that calculation will be uncertain, particularly if the total numbers are small, so results may be smoothed out too much.

The third option is to plot the curves of hazard rates, thus obtaining the best representation. Unfortunately, when behaviour patterns are compared, it is impossible to visualize similarities or differences in behaviour patterns because the plots of the hazard rate curves are mingled and chaotic.

5.4. Conclusion

In this chapter we have presented the analysis of two interacting phenomena. This method enables us to show complex dependences by developing classical demographic analysis. Unilateral dependence may be seen when a specific order can be identified, and reciprocal dependence when the two events influence each other.

As soon as the practical analysis is undertaken, the problems of simultaneity of occurrence arise. These reflect the fundamental interpretation problems raised by biographical data.

The models presented enable the data to be finely stratified, but, as we shall see in the next chapter, when the analysis is extended to cover more than two events, grouping is necessary in order to obtain statistically revealing results.

6

Extending to More Complex Situations

The bivariate situation is the simplest interaction model. The ambit of the analysis may be enlarged to cover more complex situations, such as trivariate or multivariate situations.

6.1. Presentation and Limits of the Practical Application

In this chapter we shall make a brief theoretical presentation of the problem posed by increasing the number of events studied in the same sample. We shall also see just how far the bivariate problem can go, as a result of the rapid multiplication of states.

As in the bivariate case, it is possible to consider the other events as time-dependent variables, the influence of which could be measured on the hazard rates of the event chosen; however, rather than envisage a regression study, we prefer here to analyse interactions.

Let $T_1, \ldots, T_k$, where k is the number of different events envisaged and 0 the initial state; so

$$h_{0i}(t) = \lim_{\Delta t \to 0} \frac{P(T_i < t + \Delta t \mid \bigcap_{j} \{T_j \geq t\})}{\Delta t}$$

$$h_{ij}(t \mid u) = \lim_{\Delta t \to 0} \frac{P(T_j < t + \Delta t \mid \{T_i = u\} \cap \bigcap_{l \neq i} \{T_l \geq t\})}{\Delta t} \quad \text{where } u < t$$

$$h_{lij}(t \mid u, v) = \lim_{\Delta t \to 0} \frac{P(T_j < t + \Delta t \mid \{T_i = u\} > \{T_l = v\} \cap \bigcap_{\substack{n \neq l \\ n \neq i}} \{T_n \geq t\})}{\Delta t}$$

where $u, v < t$, etc.

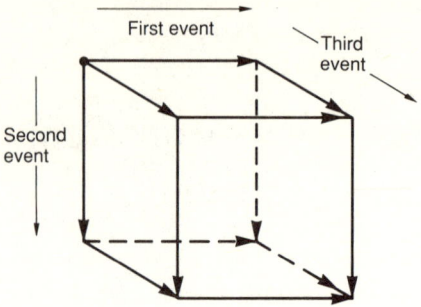

FIG. 6.1. Diagram of a trivariate study

In the bivariate case the space has four states; in the trivariate case it will have eight; and in the multivariate case with k events it will have 2^k states, which quickly becomes unenvisageable, because the data thin out as they spread out among the states.

In the trivariate case, where individuals start their progression in 0, the state-space and simple transitions (Fig. 6.1) can still be represented, without introducing the multitude of simultaneous events, and double or triple simultaneities, which would make the diagram unreadable. For a greater number of interacting events, simple representations can no longer be made.

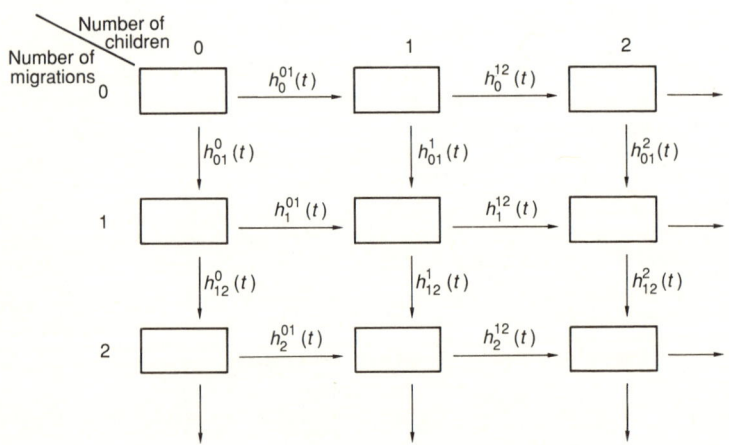

FIG. 6.2. Diagram of a multivariate two-dimensional study
Source: Courgeau and Lelièvre (1988a)

Extending to More Complex Situations 101

However, in the study of demographic events, the problem of multiplicity of states may be solved by ranking repetitive events (particularly fertility). It is, indeed, possible to conceive a study of the interactions between successive migrations and births. In this case it may be noticed that, unlike the multiple interactions observed previously, here the space is only two-dimensional (Fig. 6.2).

6.2. Interactions between Three Events: Two Study Cases

We shall now present two applications. These two examples give an idea of the method used by the researcher faced with a particular problem, and with a procedure of which he must make the best possible use.

We propose to study the processes of fertility and the separations occuring in various types of union (marriage or cohabitation). The sample studied is a set of periods of union which may or may not start by marriage and may end by a separation or a divorce. The diagram in Fig. 6.3 of a space with eight states is suitable for characterizing the situation: it should, however, be noted that some transitions are forbidden (i.e. those that are not numbered). The diagram therefore enables measurement of the following influences:

- the influence of marriage on births: comparison of (1) and (2) for a couple living together; and, reciprocally,
- the influence of a birth on marriage: comparison of (3) and (4);

(these constitute a classical bivariate case study); and, furthermore:

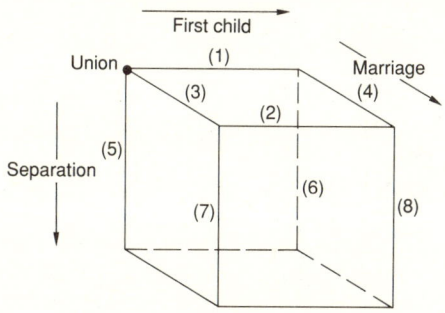

FIG. 6.3. Trivariate study of unions

102   *Extending the Scope of Longitudinal Analysis*

- the influence of a birth on the separation of a couple living together ((5), (6)) or married ((7), (8)), with no reciprocity for this, as for the following influences:
- differential separation of childless couples depending on whether they are married or living together ((7), (5));
- differential separation or not of couples having had a child depending on whether they are married or living together ((8), (6)).

However, the analysis of one particular problem is not possible with this trivariate diagram: that of the differential occurrence of births depending on whether the end of the union is a marriage or a separation, which boils down to comparing two types of direction (1) depending on the future. Thus, either a selection, or anticipation, hypothesis is tested, in which couples who are going to get married have a different fertility behaviour pattern during the period when they live together, or an adaptation hypothesis, in which all the couples that live together have the same fertility behaviour pattern, and it is only when they are married that differences arise (cf. Section 5.1.3 above).

To carry out this study, we shall take the framework of a bivariate study in which two sub-populations may be distinguished in the initial sample. Then the values of direction (1) hazard rates, obtained in the analyses on the two sub-populations, must be compared.

The second application is that of studying interactions between the first migration after marriage, the first birth, and the first change

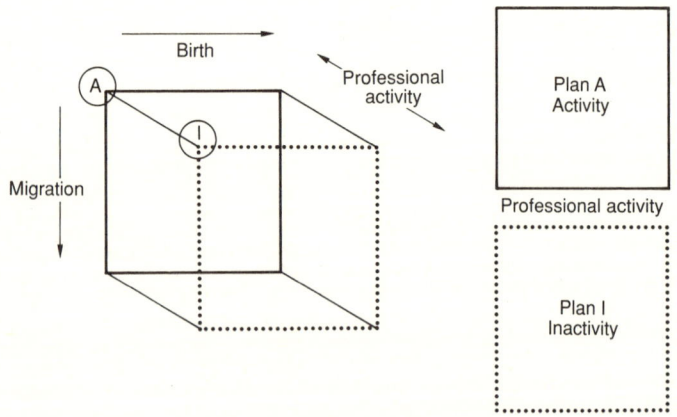

FIG. 6.4. Trivariate study: migration, birth, employment

(interruption or return) of job after marriage. This time all the transitions are envisageable. However, there is reciprocal movement between the states of job activity and inactivity (Fig. 6.4).

Up to now, the study has divided up the sample of women according to their economic activity or inactivity in marriage, and the analysis was separated for the two groups. In order to give an idea of the complexity of the calculations needed, let us count up the movements made at time t: single, double, and triple movements (simultaneous events) according to the state from which they came. In Fig. 6.5, it is possible to leave A (or I, respectively) in seven different ways.

If one of the irreversible events has already been experienced, depending on whether the subject is active or inactive at the start (four possible states, symbolized by ○ and ● in Fig. 6.6), it is possible to leave the state reached in three different ways.

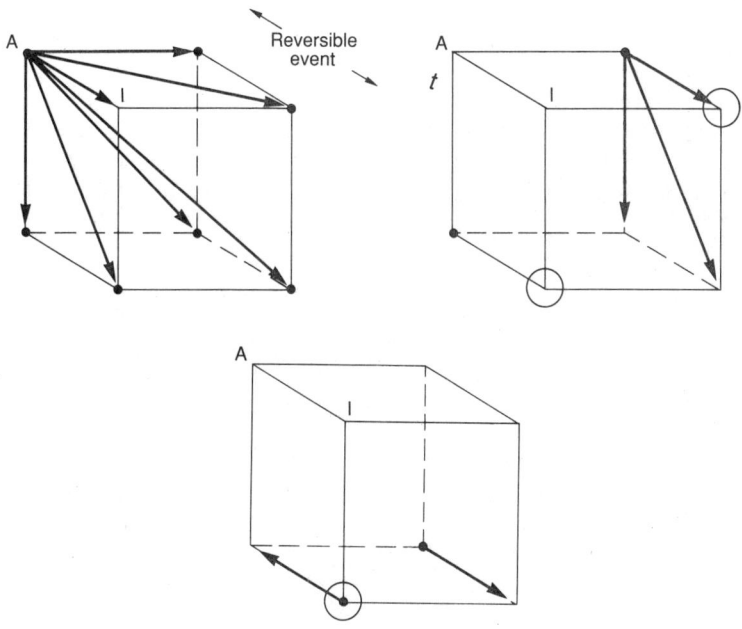

FIG. 6.5 (above left). Possible destinations after departure
FIG. 6.6 (above right). Possible directions after experiencing an irreversible event
FIG. 6.7 (opposite). Only remaining directions

Finally, after having experienced the two irreversible events, there is only one way to leave the state: through the occurrence of the reversible event that leads to the final state of the diagram (Fig. 6.7).

This application has led to non-significant comparisons from a statistical point of view, owing to the size of the initial sample. It is clear at this stage that a more formal parametric or semi-parametric modelization must be substituted for the purely non-parametric analysis. This will make it easier to discriminate between some behaviour patterns and the others, by using all the information gathered.

6.3. Interaction between Two Renewable Processes

Let us go back to the example presented in the Section 6.1, that of interaction between fertility and migrations (Courgeau, 1985*b*; Courgeau and Lelièvre, 1988*a*). The state-space (Fig. 6.2) is thus two-dimensional and each dimension is covered by random variables $T_1, T_2, \ldots, T_m, \ldots$ for the migrations, and by $T^1, T^2, \ldots, T^n, \ldots$ for the births. Each state is therefore defined by a couple (m, n) where m is the number of previous migrations and n the number of children already born. The instantaneous failure rates between the different states are of the following form:

—*for migrations*:

$$h_{i,i+1}^n(t \mid u_i, \ldots u_1, v_n, \ldots v_1)$$
$$= \lim_{\Delta t \to 0} \frac{P(T_{i+1} < t + \Delta t \mid T_{i+1} \geq t, T_i = u_i, \ldots T_1 = u_1, T^n = v_n, \ldots T^1 = v_1)}{\Delta t};$$

—*for births*:

$$h_m^{j,j+1}(t \mid u_m, \ldots u_1, v_j, \ldots v_1)$$
$$= \lim_{\Delta t \to 0} \frac{P(T^{j+1} < t + \Delta t \mid T^{j+1} \geq t, T_m = u_m, \ldots T_1 = u_1, T^j = v_j, \ldots T^1 = v_1)}{\Delta t}.$$

Clearly, such a model, imposing such a fine stratification of the initial sample into $n \times m$ states for each period, cannot be envisaged when the sample is not very large. Then, supplementary hypotheses are needed.

Extending to More Complex Situations

It may first be assumed that the intensity of the migration depends solely on the rank of this migration, on the time that has elapsed since this move (u), and on the time that has elapsed since the last birth (v). Under these hypotheses, the hazard rates thus become:

—*for migrations*:

$$h_{m,m+1}^{n}(t \mid u, v) = \lim_{\Delta t \to 0} \frac{P(T_{m+1} < t + \Delta t \mid T_{m+1} \geq t, T_m = u, T^n = v)}{\Delta t};$$

—*for births*:

$$h_{m}^{n,n+1}(t \mid u, v) = \lim_{\Delta t \to 0} \frac{P(T^{n+1} < t + \Delta t \mid T^{n+1} \geq t, T_m = u, T^n = v)}{\Delta t}.$$

These expressions are still difficult to estimate in totally non-parametric fashion, and so new hypotheses must be made. Let us admit that the intensities no longer depend on u and v, but that it is possible to study interaction between births and migrations for which only the probability of migrating while having j children is retained:

$$h^{j}(t \mid v_j) = \lim_{\Delta t \to 0} P(T_x < t + \Delta t \mid v_j \leq t, T_x \geq t),$$

and the probability of having a further child, while having already migrated i times:

$$h_i(t \mid u_i) = \lim_{\Delta t \to 0} P(T^x < t + \Delta t \mid u_i \leq t, T^x \geq t).$$

Under such hypotheses, it is possible to answer the following questions: What are the effects of n previous births on migratory processes? Reciprocally, what are the consequences of m previous migrations on fertility?

For women born between 1911 and 1925, it has thus been possible to identify different behaviour patterns depending on age at marriage. For those who married young (between 15 and 22), it is clearly shown that the size of the family has an incidence on mobility: the more births, the more mobile the family. This effect was particularly marked for the first three births. However, for women in the same cohort who married later (at over 22), this effect is non-existent. This suggests that these women had a dwelling of sufficient size for their expected families right from the start. For the two groups, reciprocally, an anticipatory effect of moving house before a birth may be detected.

These hypotheses nevertheless remain very restrictive, since they do not enable us to take into account the last residential period in a dwelling or the length of the last interval between two births, which, clearly, have a fairly important influence on estimators, particularly if the event took place in a recent time-frame. It is therefore necessary to use a more adequate modelization, taking account not only of rank, but also of the duration of stay. However, this modelization will provide less freedom, since it will be parametric or semi-parametric. This initial analysis will be extended by using the many individual characteristics available.

6.4. Conclusion

It is possible to envisage extending the capacities of the models in two directions. Either the state-space is increased, by confronting a growing number of events from distinct areas, or the interactions between two ranked series of events are studied. These generalizations to larger and larger spaces immediately introduce the problem of stratification, for no sample, even if initially exhaustive, can be subjected to this. Therefore simplifying hypotheses have to be made, which, while giving answers to questions of primary interest, nevertheless remain very strong. Any attempt to weaken them with an essentially non-parametric modelization seems practically impossible, and so recourse must be had to approaches that introduce more structure, such as semi-parametric or parametric methods. We shall investigate this interesting subject in the second part of the book.

II

Extending the Scope of Regression Models

7

Statistical Formalization of Parametric Analysis

The models presented in the first part of this work do not impose *a priori* any specific form on the hazard rates being estimated. Thus, they are non-parametric. However, it may be useful to summarize this information using a small number of parameters which make it possible to find the correct overall distribution. Apart from the saving thus made, these parameters can provide us with a simple demographic interpretation clarifying the processes observed.

We shall first present some parametric distributions chosen among those most widely used in demography. At the same time, we shall point out the underlying models which can lead to such distributions and the meaning, if any, of the estimated parameters. We shall also give some practical examples taken from different surveys to illustrate each type of model.

To take the analysis a step further, it is necessary to introduce the impact of certain individual characteristics which have been measured and may influence hazard rates. For example, an individual's educational status can have an impact on his marriage, on the birth of his children, on his migrations, or on his job mobility. These different characteristics, therefore, can be introduced as explanatory variables with regard to the individual's behaviour patterns within these parametric models. The distribution of the duration of stay will in this case depend on both qualitative and quantitative characteristics. The form that this dependence takes is discussed and various types of model are presented, along with practical examples of their application. These different characteristics make it possible to take account of the heterogeneity of the populations under study.

7.1. Some Useful Parametric Models in Demography

As before, we consider a population of individuals who are supposed to experience a given event at time T. This is a positive random variable, the distribution of which may be represented parametrically.

7.1.1. Exponential distribution

This is the simplest distribution, which is obtained when the instantaneous failure rate is a constant, independent of the duration. We may then write

$$h(t) = \rho. \tag{1}$$

The values of the survivor function can then be inferred:

$$S(t) = \exp\left(-\int_0^t h(s)ds\right) = \exp(-\rho t), \tag{2}$$

and the probability density:

$$f(t) = h(t) S(t) = \rho \exp(-\rho t), \tag{3}$$

which are both exponential functions of time. This model depends on only one parameter, ρ, which is always positive. The mean duration of stay is equal to the inverse of this parameter, $1/\rho$.

Figure 7.1 presents $S(t)$, $f(t)$, and $h(t)$ plotted against t (cf. Kalbfleisch and Prentice, 1980). This model implies a total lack of memory for the individuals at risk; whatever time period has elapsed since the initial instant, the hazard rate is independent of this duration. We can also say that, for any given moment t_0, the conditional distribution of

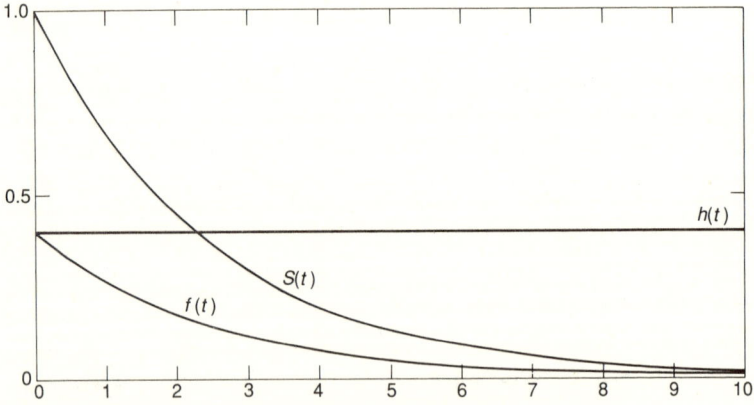

FIG. 7.1. Survivor function, $S(t)$ and probability density, $f(t)$, of an exponential distribution with constant hazard function equal to 0.4, plotted against t

$(T - t_0)$, given that T is greater than t_0, is the same as the distribution of T.

Such a distribution is observed when an individual is subjected to many reasons for experiencing the event and when he experiences the event as soon as one of these reasons occurs. Let us examine in more detail how to represent this mechanism.

Let $T_1, \ldots, T_n$ be independent random instants having the same distribution according to time. This distribution is such that when $t \to 0$ one can write

$$S(t) \simeq 1 - \rho t$$

$$f(t) \simeq \rho. \tag{4}$$

If we denote as M_n the minimum of the random instants $T_1, \ldots, T_n$, then the density of M_n can be written

$$f_{M_n}(t) = n\rho(1 - \rho t)^{n-1}, \tag{5}$$

since this minimum is close to value 0. It follows that the random variable $X_n = nM_n$ has a density

$$f_{X_n}(t) = \lim_{\Delta t \to 0} \frac{P(t \leq nM_n < t + \Delta t)}{\Delta t}$$

$$= \frac{1}{n} \lim_{\Delta t/n \to 0} \frac{P\left(\frac{t}{n} \leq M_n < \frac{t}{n} + \frac{\Delta t}{n}\right)}{\frac{\Delta t}{n}} = \rho\left(1 - \frac{\rho t}{n}\right)^{n-1} \tag{6}$$

In this case the survivor function is written

$$S_{X_n}(t) = \left(1 - \frac{\rho t}{n}\right)^n, \tag{7}$$

and the hazard rate,

$$h_{X_n} = \frac{\rho}{1 - \frac{\rho t}{n}} \tag{8}$$

Thus, when $n \to \infty$, $h_{X_n}(t)$ really tends towards the constant ρ.

Let us now see how to check empirically whether this distribution can be used. Various simple tests may be carried out.

112   *Extending the Scope of Regression Models*

Using the cumulative hazard rate, rather than the instantaneous failure rate, which can present large variations, we can write

$$H(t) = \int_0^t h(s)ds = \rho t. \qquad (9)$$

Thus, plotting the cumulative hazard rate against t leads to a straight line, if the exponential model is appropriate to a set of data.

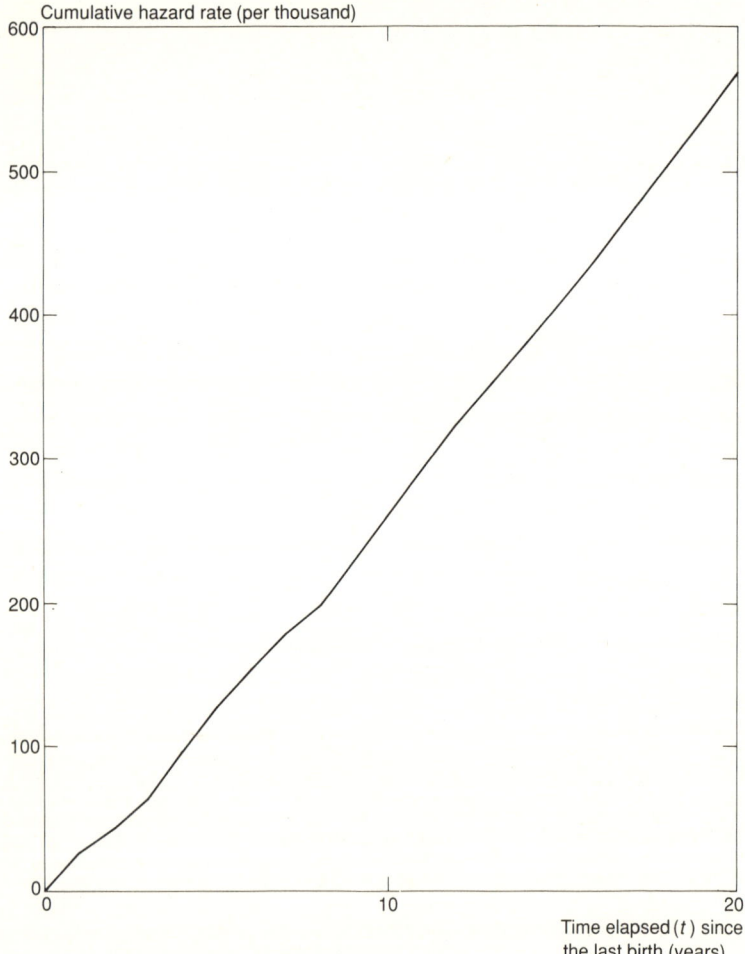

FIG. 7.2. Cumulative hazard rates of women who have become home-owners after their last child's birth, plotted against the duration since that birth

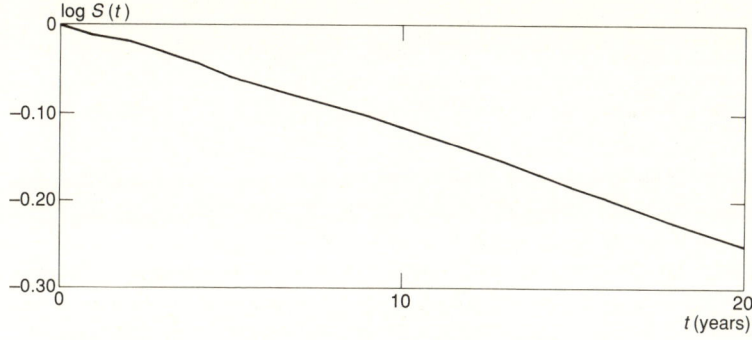

FIG. 7.3. Logarithm of the survivor function of women not yet home-owners since their last child's birth, according to the time elapsed since that birth

Figure 7.2 gives an example of the use of this method to study the probability of becoming a home-owner after the last child's birth, obtained with the 'Triple Biographie' survey data. The durations are computed for complete families, starting from the last child's birth. Women who have had no children are therefore not included in the sample. Hazard rates are estimated using the non-parametric methods described in the first part of the book. This figure, which plots the cumulative hazard rates, shows that these results can be represented satisfactorily by using an exponential distribution.

Another empirical check of the appropriateness of the exponential model uses the logarithm of the survivor function. If the exponential model fits the data, we can in fact write

$$\log S(t) = -\rho t. \qquad (10)$$

It follows that the logarithm of the survivor function plotted against t, if the model fits the data, should approximate a straight line passing through the origin. Figure 7.3 reports those results for the same data as in Fig. 7.2. Again, it appears that these data can be represented with an exponential distribution.

7.1.2. Mixing exponential distributions

In the previous case it was assumed that the population is homogeneous: all individuals have the same probability of experiencing the event. This hypothesis is hardly likely to be verified. It may be assumed that the population breaks down into various sub-populations, each having a time-independent hazard rate. The most general case is that where each individual has a specific failure rate, but always independent of time.

We shall proceed from the most simple case, where the population is divided into two sub-populations, to the most complex case, where each individual has a different failure rate.

In the first case, suppose that the first sub-population is not at risk, whereas the second is. The model is then called a 'mover–stayer model', frequently used in the case of internal migrations (Courgeau, 1973) or job mobility (Blumen et al., 1955). This model has already been presented briefly in Section 2.2.2.

Let us denote $S(\infty)$ as the probability of never experiencing the event. The survivor function of the second sub-population, which is at risk, is then the part $(S(t) - S(\infty))$, at time t. If the hazard rate of this sub-population is equal to ρ, then we can write in a differential form

$$d(S(t) - S(\infty)) = -\rho(S(t) - S(\infty))dt, \qquad (11)$$

and hence

$$S(t) = S(\infty) + (1 - S(\infty)) \exp(-\rho t). \qquad (12)$$

The probability density function is equal to the opposite of the derivated function of $S(t)$:

$$f(t) = \rho(1 - S(\infty)) \exp(-\rho t), \qquad (13)$$

and the hazard rate is written

$$h(t) = \frac{\rho(1 - S(\infty)) \exp(-\rho t)}{S(\infty) + (1 - S(\infty)) \exp(-\rho t)} = \frac{\rho}{1 + \dfrac{S(\infty)}{1 - S(\infty)} \exp(\rho t)}. \qquad (14)$$

Then, the population as a whole no longer has a constant hazard rate, unlike the two sub-populations. When the duration is short, this rate is close to the ρ value, which is the rate for the mobile population. When this duration increases, the rate tends towards zero, which is the rate for the immobile population. The observed variations of the rate for the whole population follow the variations of the composition of the population, which is different at the beginning and at the end of the observation period.

Formalization of Parametric Analysis

The mean duration of stay in the initial state for the second sub-population will be

$$\int_0^\infty \rho t(1 - S(\infty)) \exp(-\rho t) dt = \frac{(1 - S(\infty))}{\rho}.\qquad(15)$$

Figure 7.4 indicates the various distributions versus time t when the mover–stayer model is verified. They are all monotone-decreasing.

Let us now determine how to verify empirically the fit of this model. If the annual variation of the survivor function is examined, we can write

$$S(t-1) - S(t) = (1 - S(\infty))(\exp \rho - 1)\exp(-\rho t).\qquad(16)$$

When proceeding to logarithms, this gives

$$\log \Delta S(t) = \log c - \rho t,\qquad(17)$$

where $\Delta S(t) = S(t-1) - S(t)$ and $c = (1 - S(\infty))(\exp \rho - 1)$.

By reporting on a semi-logarithmic graded paper the values of $\log \Delta S(t)$ versus t, we obtain a straight line of slope, $-\rho$, when the mover–stayer model fits the data.

Figure 7.5 reports these values, for changes of dwelling of women born between 1926 and 1936, according to the duration of stay. The data are taken from the 'Triple Biographie' survey. These points

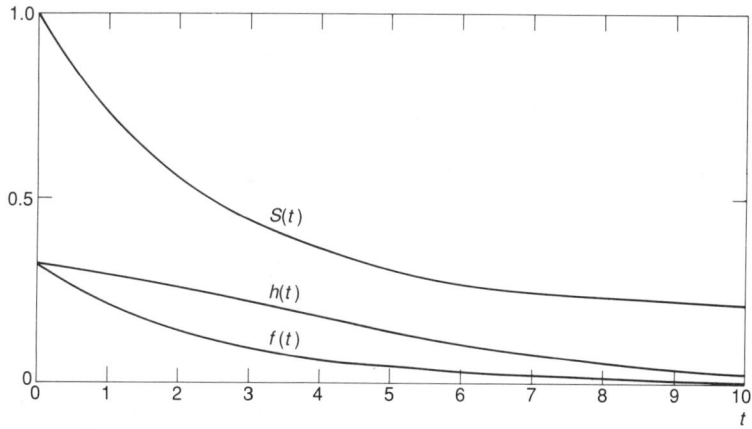

FIG. 7.4. Duration of stay, $S(t)$, probability density, $f(t)$, and hazard rate, $h(t)$, of a mover–stayer model with parameter $\rho = 0.4$ and $S(\infty) = 0.2$, plotted against time t

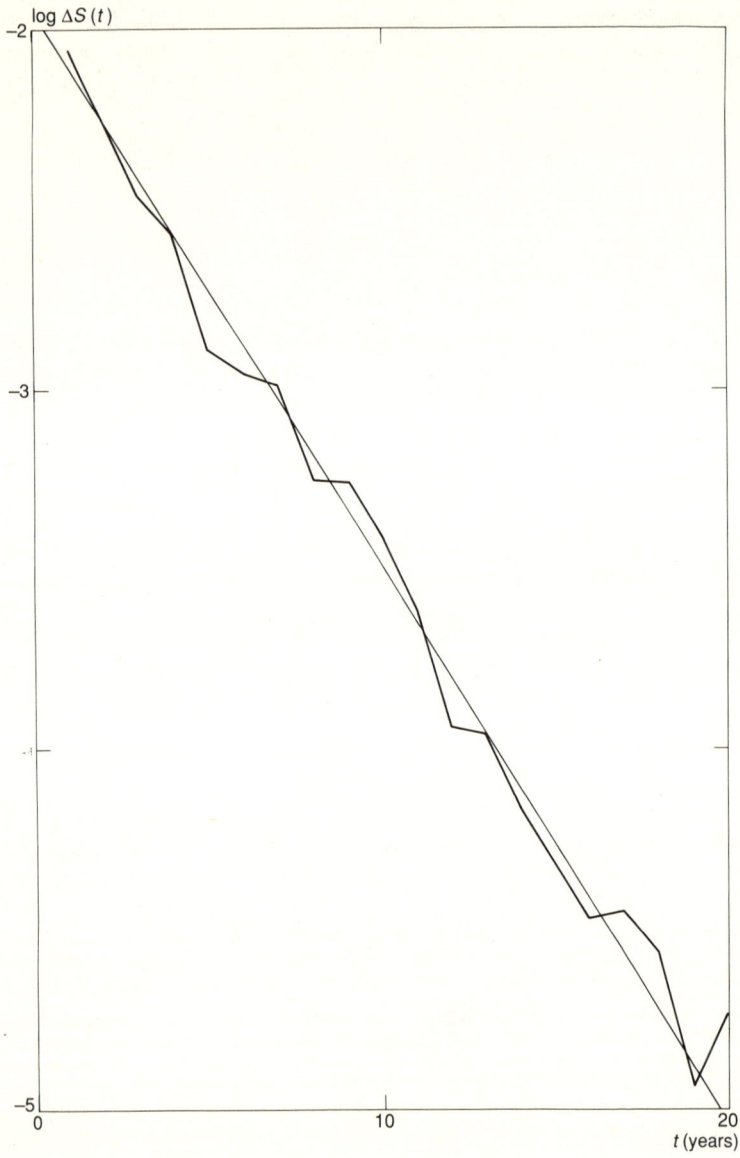

FIG. 7.5. Logarithm of the survivor function for changes of dwelling of women born between 1926 and 1936

Formalization of Parametric Analysis 117

concern non-aggregated data and are more dispersed than in the previous examples. Nevertheless, they are correctly aligned and the mover–stayer model is verified in this case. Migrants have a mobility rate of about 0.15 and they represent 86 per cent of the initial population.

This model can be made more complex by introducing two migration probabilities that are not nil and are different for each sub-population, or by introducing n sub-populations, each having a different probability of migration. These models are treated similarly to the mover–stayer model.

Let us examine the most complex case, where there is a mix of an infinite number of exponential distributions. In that case each individual still has a constant hazard rate, but this rate varies from one individual to another. This makes it possible to introduce a heterogeneity between individuals.

Let us define in this case a new random variable P, with density $f_P(\cdot)$, which stands for the distribution of the various rates in the population. The conditional density of T, given $P = \rho$ for a given individual, is as before:

$$f_{T|P}(t \mid \rho) = \rho \exp(-\rho t). \tag{18}$$

In that case, the unconditional density by P is obtained by summing up on all the possible values of ρ:

$$f_T(t) = \int_0^\infty \rho \exp(-\rho t) f_P(\rho) d\rho. \tag{19}$$

The survivor function is then written

$$S_T(t) = \int_t^\infty f_T(\theta) d\theta = \int_{\theta=t}^\infty \int_{\rho=0}^\infty \rho \exp(-\rho\theta) f_P(\rho) d\rho d\theta.$$

Hence

$$S_T(t) = \int_{\rho=0}^\infty \rho f_P(\rho) d\rho \int_{\theta=t}^\infty \exp(-\rho\theta) d\theta = \int_{\rho=0}^\infty \exp(-\rho t) f_P(\rho) d\rho. \tag{20}$$

Note that the survivor function is the unilateral Laplace transform of the probability density $f_P(\rho)$ of variable P. We know that in this case, whatever the probability density of variable P, the density and the unconditional survivor function are always monotonous functions of t.

According to the probability density of variable P, different types of density can be obtained for variable T. If, for instance, variable P is exponentially distributed, with mean value ρ_0;

$$f_P(\rho) = \frac{1}{\rho_0} \exp\left(-\frac{\rho}{\rho_0}\right), \quad (21)$$

it then follows that

$$f_T(t) = \int_0^\infty \frac{\rho}{\rho_0} \exp\left[-\rho\left(t + \frac{1}{\rho_0}\right)\right] d\rho = \frac{1}{\rho_0\left(t + \frac{1}{\rho_0}\right)^2}. \quad (22)$$

Hence

$$S_T(t) = \frac{1}{\rho_0 t + 1} \quad (23)$$

and

$$h_T(t) = \frac{\rho_0}{\rho_0 t + 1}. \quad (24)$$

Note that the density, the survivor function, and the hazard rate follow a Pareto distribution. At the initial instant, the hazard rate is equal to the mean hazard rate of the population, ρ_0. Figure 7.6 reports these distributions when $\rho_0 = 0.4$.

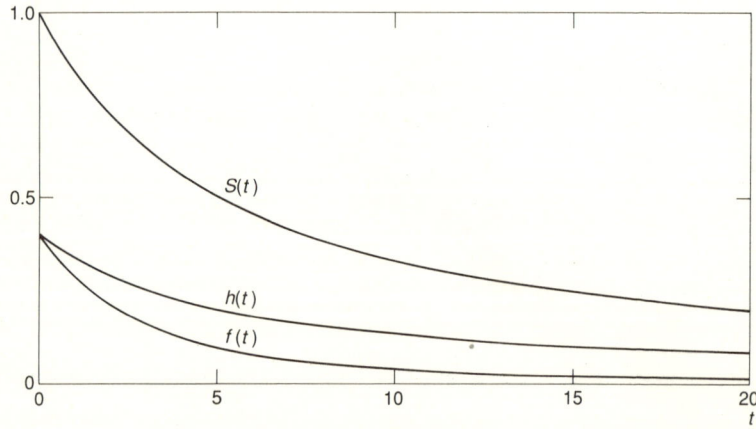

FIG. 7.6. Survivor function, $S(t)$, probability density, $f(t)$, and hazard rate, $h(t)$, of a Pareto distribution of parameter $\rho = 0.4$

Formalization of Parametric Analysis

More generally, if the probability density of P is a gamma distribution, of mean ρ_0, we can then write

$$f_P(\rho) = \frac{\lambda}{\rho_0}\left(\frac{\lambda\rho}{\rho_0}\right)^{\lambda-1} \frac{\exp\left(-\frac{\lambda\rho}{\rho_0}\right)}{\Gamma(\lambda)}, \qquad (25)$$

where

$$\Gamma(\lambda) = \int_0^\infty x^{\lambda-1} \exp(-x)\,dx. \qquad (26)$$

We can show that here, too, the density follows a Pareto distribution:

$$f_T(t) = \lambda \left(\frac{\lambda}{\rho_0}\right)^{\lambda} \left(t + \frac{\lambda}{\rho_0}\right)^{-(\lambda+1)}. \qquad (27)$$

The survivor function is then equal to

$$S_T(t) = \left(\frac{\lambda}{\rho_0}\right)^{\lambda} \left(t + \frac{\lambda}{\rho_0}\right)^{-\lambda}, \qquad (28)$$

and the hazard rate to

$$h_T(t) = \frac{\lambda}{t + \frac{\lambda}{\rho_0}}. \qquad (29)$$

Again, these three functions are Pareto distributions. The density at the initial instant is equal to the average value for the whole population, ρ_0. If $\lambda = 1$, the case is the same as before, since the exponential distribution is a particular case of a gamma distribution.

It is clear, then, that the Pareto distribution appears when the population under study can be considered to be composed of individuals with exponential survivor function, who are themselves distributed following a gamma distribution.

We can verify empirically if this distribution is satisfactory by plotting the inverse of the hazard rate values against t. In fact, formulas (24) and (29) give

$$\frac{1}{h_T(t)} = \frac{t}{\lambda} + \frac{1}{\rho_0}. \qquad (30)$$

If this relation is verified, then the curve is linear.

Figure 7.7 plots these values for the changes of dwelling of women born between 1926 and 1936, statistics already used to test the validity of the mover–stayer model. The relation is hardly linear, and we cannot make the hypothesis of an exponential or gamma distribution of P, whereas the mover–stayer model fits this distribution perfectly.

Obviously, we can envisage many other types of distribution of P, which will lead to many different distributions of the densities, survivor functions, and hazards observed in the whole population. It

FIG. 7.7. Inverse of the hazard rate for changes of dwelling of women born between 1926 and 1936, in relation to the survivor function

must be remembered that these distributions will always remain monotonous time functions.

7.1.3. Gompertz distribution

Let us consider again a homogeneous population. If the variation of the hazard rate over time is proportional to its value at each instant, then a Gompertz distribution is observed. This condition can be written

$$\frac{dh(t)}{dt} = \rho h(t), \tag{31}$$

which gives, once integrated,

$$h(t) = \lambda \rho \exp(\rho t). \tag{32}$$

According to parameter ρ, the rate is either monotone-increasing ($\rho > 0$) or monotone-decreasing ($\rho < 0$). When $\rho = 0$ and $\lambda \rho$ is reduced to a constant, we have the same exponential distribution as before. Parameter λ is always of sign ρ, so that the rate should be positive.

When coefficient ρ is positive, this distribution does apply to data on mortality, at least for ages over 35. It has also been applied to data on migration or occupational mobility, but with a negative coefficient. Figure 7.8 indicates the distribution of the rate according to time for several values of parameter ρ, which enables us to make a clear distinction between the cases.

The cumulative hazard rate can be written

$$H(t) = \int_0^t \lambda \rho \exp(\rho \theta) d\theta = \lambda [\exp(\rho t) - 1]. \tag{33}$$

Hence

$$S(t) = \exp(\lambda [1 - \exp(\rho t)]). \tag{34}$$

It appears therefore that, when parameter ρ is negative, the value of parameter λ can be interpreted by writing, when $t \to \infty$,

$$S(\infty) = \exp \lambda. \tag{35}$$

In this case the model is similar to the mover–stayer model, since part of the population never experiences the event. We can then write

$$S(t) = S(\infty) \exp[-\log S(\infty) \exp \rho t] \quad \text{if } \rho < 0, \tag{36}$$

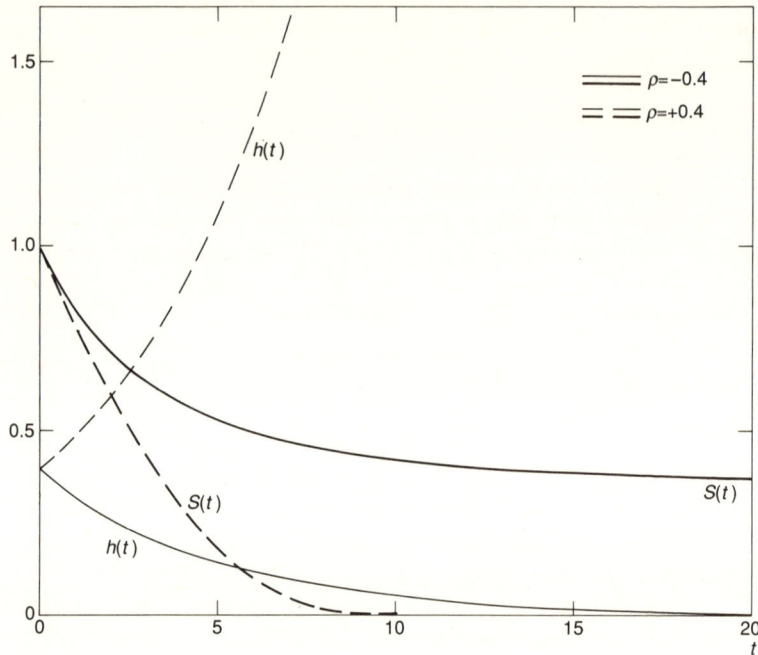

FIG. 7.8. Survivor functions, $S(t)$, and instantaneous failure rates, $h(t)$, with two distributions of parameters ($\rho = -0.4, \lambda = -1$) and ($\rho = 0.4, \lambda = 1$)

which gives, when differentiating,

$$d(\log S(t) - \log S(\infty)) = \rho(\log S(t) - \log S(\infty))dt. \qquad (37)$$

It appears that the previous relation (11) is not verified with $S(t)$ but is verified with $\log S(t)$.

Considering again the general model, the density function can be written

$$f(t) = \lambda\rho \exp[\rho t + \lambda(1 - \exp(\rho t))]. \qquad (38)$$

This density can also be written

$$f(t) = \rho \exp(\lambda) \exp[\rho t + \log(\lambda) - \exp(\rho t + \log(\lambda))]. \qquad (39)$$

To check empirically whether this distribution can be used, we can plot the logarithm of $h(t)$ against time t. If we obtain an approximation of a straight line, it means that the distribution can be used.

Formalization of Parametric Analysis

In Fig. 7.9 we have depicted these values for the changes of dwelling of women born between 1926 and 1936 in relation to the duration of stay. The data are the same as those used for testing the mover–stayer model. *A priori*, the Pareto model is equally satisfactory for these data. A more precise testing is then necessary in order to choose between the two models.

It is also possible to work on the annual variation of the logarithm of the survivor function. In fact, we can write

$$\log S(t-1) - \log S(t) = \lambda(1 - \exp(-\rho))\exp(\rho t), \qquad (40)$$

which gives, when proceeding to logarithms,

$$\log(\Delta \log S(t)) = \log c + \rho t, \qquad (41)$$

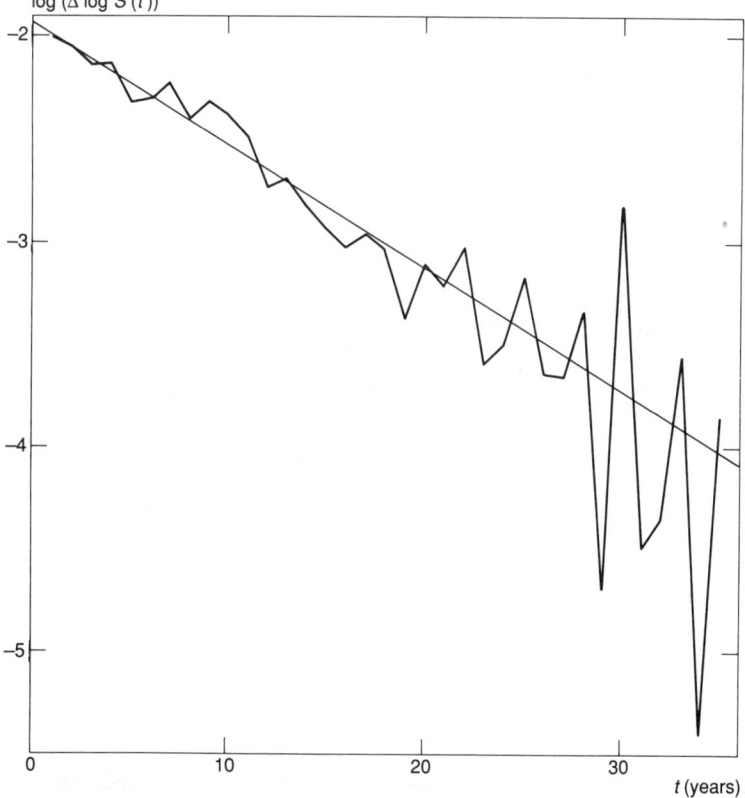

FIG. 7.9. Logarithm of the hazard rate in relation to the duration of stay for changes of dwelling of women born between 1926 and 1936

where

$$\Delta \log S(t) = \log S(t-1) - \log S(t) \quad \text{and} \quad c = \log [\lambda (1 - \exp(-\rho))].$$

Plotting the values of $\log (\Delta \log S(t))$ against time t results in a line, if the Gompertz model is verified.

Figure 7.10 reports these values for the changes of dwelling of women born between 1926 and 1935 in relation to the duration of stay. Again, it shows that these migrations can be approached through a Gompertz model.

It is possible to introduce another parameter in the model, thus obtaining a Gompertz–Makeham model, for which the hazard rate is written

$$h(t) = \rho_0 + \lambda \rho \exp(\rho t), \tag{42}$$

which gives

$$\log S(t) = -\rho_0 t + \lambda [1 - \exp(\rho t)] \tag{43}$$

and

$$f(t) = [\rho_0 + \lambda \rho \exp(\rho t)] \exp [-\rho_0 t + \lambda (1 - \exp(\rho t))]. \tag{44}$$

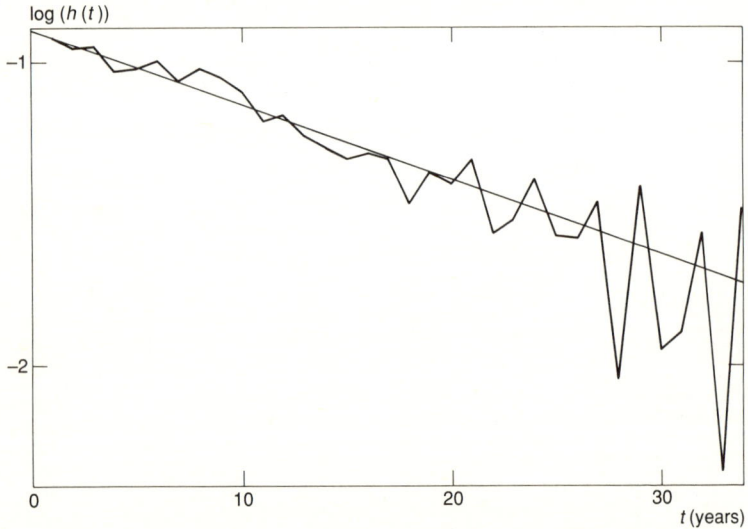

FIG. 7.10. Logarithm of the annual variation of the logarithm of the survivor function, $S(t)$, in relation to duration of stay, for changes of dwelling of women born between 1926 and 1936

This model, which introduces one more parameter than the Gompertz model, generally fits the empirical data better.

7.1.4. Weibull distribution

The Weibull distribution provides another type of monotonous time function for the rate. The rate varies as a given exponent of time, which is written

$$h(t) = \lambda \rho (\rho t)^{\lambda - 1}, \qquad (45)$$

where λ and ρ are positive parameters. When $\lambda = 1$, we again have an exponential distribution. When λ is greater than the unit, the hazard rate is monotone-increasing; when λ is less than the unit, it is monotone-decreasing.

Figure 7.11 indicates this rate in relation to time for several values of λ and of ρ. By integrating in relation to time, it appears that the cumulative hazard rate is written

$$H(t) = \int_0^t \lambda \rho (\rho \theta)^{\lambda - 1} \, d\theta = (\rho t)^\lambda. \qquad (46)$$

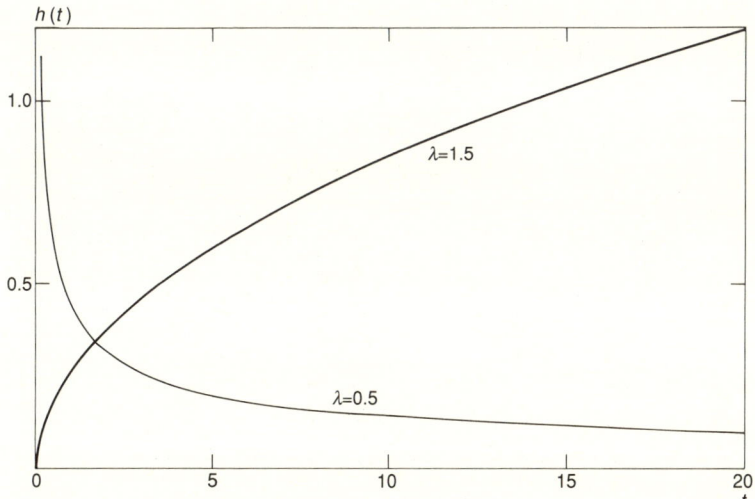

FIG. 7.11. Instantaneous failure rates of two Weibull distributions in which the parameter ρ is equal to 0.4 and the λ parameters are equal to 0.5 and 1.5 respectively

Hence
$$S(t) = \exp[-(\rho t)^\lambda] \quad (47)$$

and
$$f(t) = \lambda \rho (\rho t)^{\lambda-1} \exp[-(\rho t)^\lambda]. \quad (48)$$

This distribution, like the exponential distribution, is observed when there are many reasons for an individual to experience the event observed and to experience it as soon as one of the reasons occurs.

Then let $T_1, \ldots, T_n$ be independent random instants with the same distribution in relation to time. This distribution is such that when $t \to 0$ we can write

$$S(t) \simeq 1 - (\rho t)^\lambda$$
$$f(t) \simeq \lambda \rho (\rho t)^{\lambda-1}. \quad (49)$$

If M_n denotes the minimum of the $T_1, \ldots, T_n$ random instants, we can write the density of M_n, as in the exponential case:

$$f_{M_n}(t) = n\lambda \rho (\rho t)^{\lambda-1} [1-(\rho t)^\lambda]^{n-1}. \quad (50)$$

This results in the random variable $X_n = n^{1/\lambda} M_n$ having a density equal to

$$f_{X_n}(t) = \lim_{\Delta t \to 0} \frac{P(t \leq n^{1/\lambda} M_n < t + \Delta t)}{\Delta t}$$
$$= n^{-1/\lambda} n\lambda \rho (\rho t)^{\lambda-1} n^{(1-\lambda)/\lambda} [1-(\rho t)^\lambda n^{-1}]^{n-1}$$
$$= \lambda \rho (\rho t)^{\lambda-1} [1-(\rho t)^\lambda n^{-1}]^{n-1}. \quad (51)$$

If follows that
$$S_{X_n}(t) = [1-(\rho t)^\lambda n^{-1}]^n \quad (52)$$

and
$$h_{X_n}(t) = \frac{\lambda \rho (\rho t)^{\lambda-1}}{1-(\rho t)^\lambda n^{-1}}. \quad (53)$$

Therefore, when $n \to \infty$, the rate approaches the value

$$h_{X_n}(t) = \lambda \rho (\rho t)^{\lambda-1},$$

which is truly a Weibull distribution, already given in the previous formula (45).

Formalization of Parametric Analysis

To verify empirically whether a Weibull model can be used, we test the values of $H(t)$, or of $-\log S(t)$. We thus write

$$\log [H(t)] = \log [-\log S(t)] = \lambda (\log \rho + \log t). \tag{54}$$

Let us indicate on a graph the logarithm of the opposite of the logarithm of the survivor function in relation to the logarithm of time t. When the Weibull model can be used, we obtain a line with a slope of λ and an intersection with the times axis which gives an estimate of $-\log \rho$.

Figure 7.12 reports the results obtained by working on the changes of dwelling of women born between 1926 and 1935. Again, it appears that this distribution can be approached by a Weibull distribution where the approximate value of parameters is $\lambda = 0.84$ and $\log \rho = 1.3$. However, the adjustment does not seem to be as good as with a mover–stayer or a Gompertz model. Below, we shall see how to choose between these distributions.

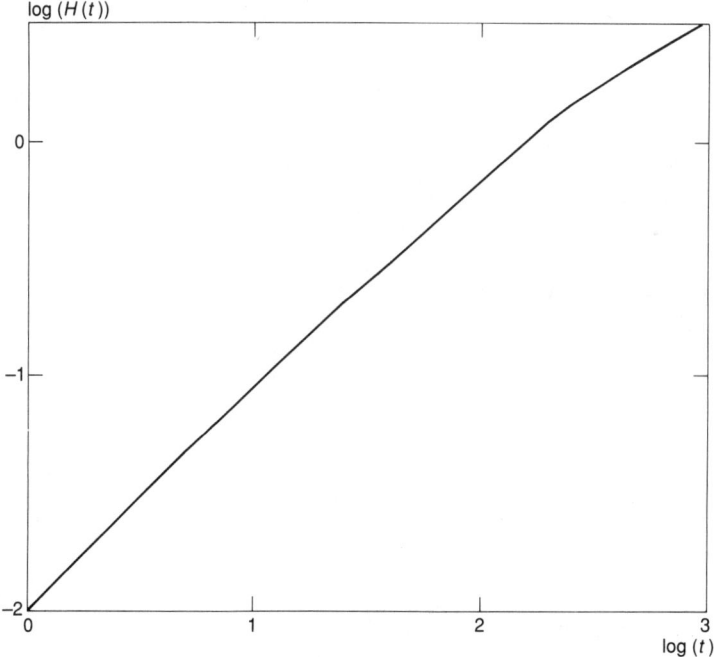

FIG. 7.12. Logarithm of the cumulative hazard rate, $H(t)$, in relation to the logarithm of the duration of stay in their dwelling of women born between 1926 and 1936

7.1.5. Gamma distribution

This is the last type of monotonous time function for the rate that we present here.

This distribution is obtained when an individual is subjected to a certain number of risks, all distributed exponentially, with the same parameter ρ. The individual finally experiences the event studied when a given number, λ, of these risks have occurred. In that case, the probability density of the event is

$$f(t) = \frac{\rho(\rho t)^{\lambda-1} \exp(-\rho t)}{(\lambda-1)!}. \tag{55}$$

This distribution can be generalized when λ is not an integer. We thus write

$$f(t) = \frac{\rho(\rho t)^{\lambda-1} \exp(-\rho t)}{\Gamma(\lambda)}, \tag{56}$$

where the $\Gamma(\lambda)$ function is defined by the following relation:

$$\Gamma(\lambda) = \int_0^\infty x^{\lambda-1} \exp(-x) dx, \tag{57}$$

where $\lambda > 0$.

Although this distribution is among the most used in statistics among continuous distributions defined for $t \geq 0$, it is more difficult to use in event history analysis because of the complexity of the survivor function and hazard rates.

In fact, we can write

$$S(t) = 1 - \frac{\int_0^t x^{\lambda-1} \exp x \, dx}{\Gamma(\lambda)} \tag{58}$$

and

$$h(t) = \frac{\rho(\rho t)^{\lambda-1} \exp(-\rho t)}{\Gamma(\lambda) - \int_0^t x^{\lambda-1} \exp x \, dx} \tag{59}$$

The hazard rate increases from value 0 to value λ when $\lambda > 1$, and decreases from the infinite to λ when $\lambda < 1$. When $\lambda = 1$, we again have an exponential model.

7.1.6. Log-normal distribution

We shall now examine distributions of which the rates are no longer monotone-increasing or -decreasing functions of time. One possibility consists in considering the variable $Y = \log T$ as normally distributed, of mean $\log 1/\rho$ and of standard deviation σ. In that case, T has a log-normal distribution with a density written

$$f(t) = \frac{1}{\sigma t \sqrt{2\pi}} \exp\left(-\frac{[\log(t\rho)]^2}{2\sigma^2}\right). \tag{60}$$

Computing the survivor function and the hazard rate involves using the incomplete normal integral:

$$\varphi(x) = \frac{1}{\sqrt{2\pi}} \int_{-\infty}^{x} \exp\left(-\frac{u^2}{2}\right) du, \tag{61}$$

which gives as value for the survivor function

$$S(t) = 1 - \varphi\left[\frac{1}{\sigma} \log(t\rho)\right] \tag{62}$$

and as hazard rate

$$h(t) = \frac{1}{\sigma t \left(1 - \varphi\left[\frac{1}{\sigma} \log(t\rho)\right]\right) \sqrt{2\pi}} \exp\left(-\frac{[\log(t\rho)]^2}{2\sigma^2}\right). \tag{63}$$

In Fig. 7.13 we have indicated the hazard rates obtained for various values of σ. These rates are no longer monotonous time functions, but they go through a maximum.

Since the survivor function makes it necessary to use the normal incomplete integral, it gives rise to problems of estimation which are difficult to solve in a simple way, when some of the intervals are truncated. In this case, it is better to use the log-logistic distribution, which provides a correct approximation of the log-normal distribution. Its duration of stay or its hazard rate can also be expressed in a simpler way in relation to time.

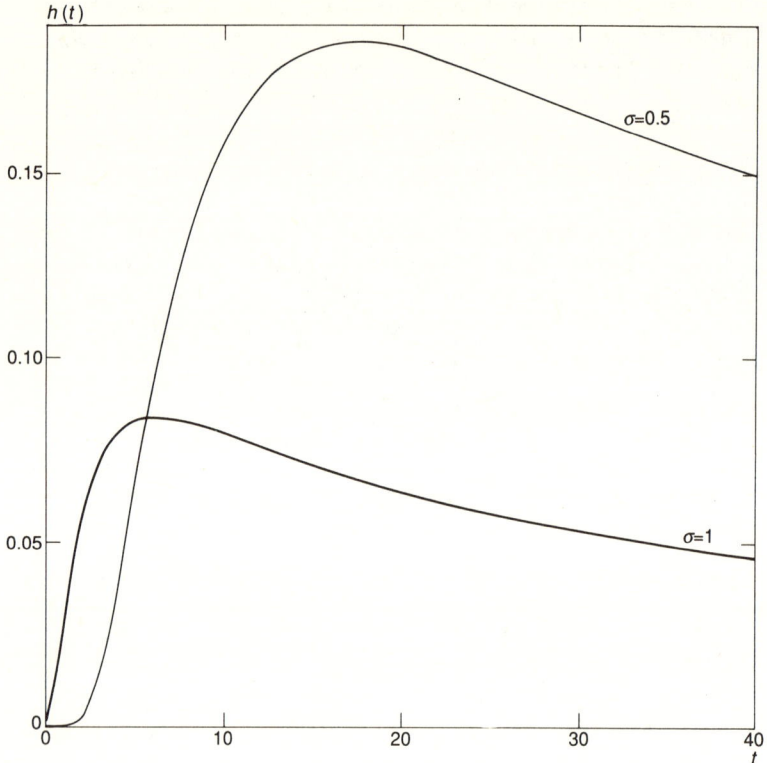

FIG. 7.13. Instantaneous failure rates of two log-normal distributions where parameter ρ equals 0.1 and parameters σ equal 0.5 and 1 respectively

7.1.7. Log-logistic distribution

We know that the form of a logistic distribution is very similar to that of a normal distribution, to the extent that we can often substitute one for the other without introducing any significant bias.

By

$$\log T = Y = -\log \rho + \sigma W, \tag{64}$$

we obtain a log-logistic distribution for T when W has a logistic distribution:

$$f_W(w) = \exp w [1 + \exp w]^{-2}. \tag{65}$$

This implies that the probability density of T, by $\lambda = 1/\sigma$, is

$$f(t) = \frac{\lambda \rho (\rho t)^{\lambda - 1}}{[1 + (\rho t)^\lambda]^2} \qquad (66)$$

the survivor function is,

$$S(t) = [1 + (\rho t)^\lambda]^{-1}, \qquad (67)$$

and the hazard rate is

$$h(t) = \lambda \rho (\rho t)^{\lambda - 1} [1 + (\rho t)^\lambda]^{-1}. \qquad (68)$$

This distribution is advantageous compared with log-normal distributions in that the survivor function and the hazard rate have simple explicit forms. When data are partly censored, the model parameters can easily be interpreted.

Figure 7.14 indicates the distribution of the hazard rates for different values of λ. If $\lambda > 1$, then the hazard rates curve goes through a maximum at $t = (\lambda - 1)^{1/\lambda}/\rho$; if $\lambda < 1$, it is constantly decreasing.

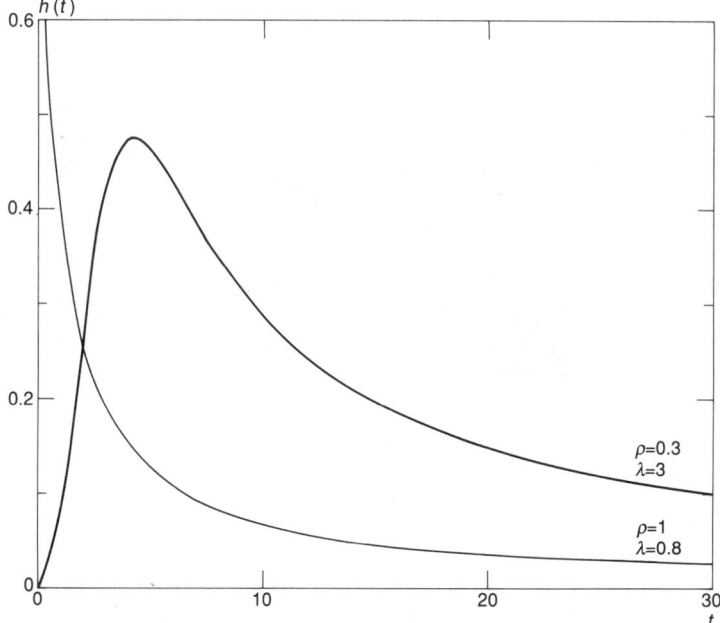

FIG. 7.14. Hazard rates of two log-logistic distributions with parameters of ($\rho = 1$, $\lambda = 0.8$) and ($\rho = 0.3$, $\lambda = 3$) respectively

Note that, if the log-logistic distribution fits the data, then

$$\frac{S(t)}{1-S(t)} = (\rho t)^{-\lambda}. \tag{69}$$

The logarithm of the probability of experiencing the event after t divided by the probability of experiencing it before t is then a linear function of the logarithm of t.

It appears that the hazard rate can still be written in the following form:

$$h(t) = \frac{\lambda}{t}[1 - S(t)]. \tag{70}$$

This relation means that the hazard rate is proportional to the number of individuals who have already experienced the event, and inversely proportional to the duration of stay.

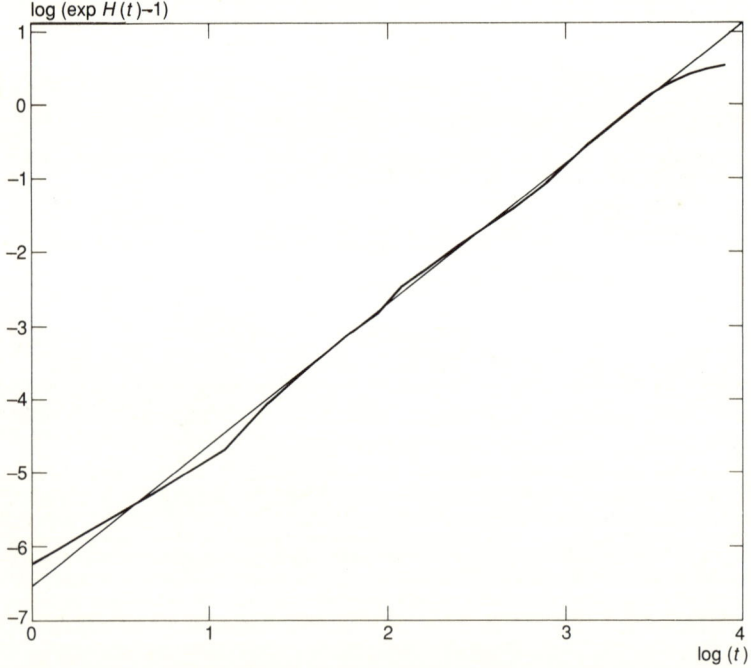

FIG. 7.15. Logarithm of the inverse of the survivor function minus the unit plotted against the logarithm of t, with regard to the relationship between becoming a home-owner and the age of individuals from 15 onwards

To check the appropriateness of this model, we can write

$$\log\left[\frac{1-S(t)}{S(t)}\right] = \log(\exp H(t) - 1) = \lambda \log(\rho t). \tag{71}$$

If the model accurately fits the observation, a line can be obtained by plotting the logarithm of the inverse of the survivor function minus the unit against the logarithm of the duration. These values are reported in Fig. 7.15, with regard to the relationship between becoming a home-owner and the age of individuals from 15 onwards. The distribution of these data can accurately be described by a log-logistic distribution, at least until age 50.

7.1.8. Generalized Fisher–Snedecor (F) distribution

This last parametric model incorporates most of the distributions hitherto presented, as well as particular cases. It should therefore make it possible to choose, from among all the distributions, the one that fits the data best, as will be shown in the following chapter.

Let us again assume

$$\log T = Y = -\log \rho + \sigma W, \tag{72}$$

where the density probability of W is a distribution equal to

$$f_W(w) = \frac{k_1^{k_1} k_2^{k_2} \Gamma(k_1 + k_2) \exp(k_1 w)}{\Gamma(k_1) \Gamma(k_2) (k_1 + k_2 \exp w)^{k_1 + k_2}}, \tag{73}$$

where $2k_1$ and $2k_2$ are the degrees of freedom of this distribution and $\Gamma(k)$ is the function Γ already defined. If $k_1 = k_2 = 1$, for instance, the model becomes a logistic distribution:

$$f_W(w) = \frac{\exp(w)}{(1 + \exp w)^2}, \tag{74}$$

so that there is a log-logistic distribution for T.

Other types of models each correspond to particular values of k_1 and k_2, which can tend towards the infinite. Thus, for example, when $k_1 = 1$ and $k_2 \to \infty$, we can write

$$\lim_{k_2 \to \infty,\ k_1 = 1} f_W(w) = \exp(w - \exp w). \tag{75}$$

In that case, a Weibull distribution is observed when $\sigma \neq 1$, and an exponential one when $\sigma = 1$. (For other cases, see Kalbfleish and Prentice, 1980: 28–9).

Using relation (81), it is possible to compute the probability density of T, which is equal to

$$f(t) = \frac{k_1^{k_1} k_2^{k_2} \Gamma(k_1 + k_2) \rho (\rho t)^{(k_1/\sigma) - 1}}{\Gamma(k_1) \Gamma(k_2) \sigma [k_2 + k_1 (\rho t)^{(1/\sigma)}]^{k_1 + k_2}}. \qquad (76)$$

Again, if $k_1 = k_2 = 1$, and assuming $\lambda = 1/\sigma$, T has a log-logistic distribution, then

$$f(t) = \frac{\lambda \rho (\rho t)^{\lambda - 1}}{[1 + (\rho t)^{\lambda}]^2}. \qquad (77)$$

7.1.9. Comparison of various distributions

Table 7.1 summarizes the main properties that make it possible to choose from among several distributions. These characteristics remain valid when working on data that include censored intervals.

Using cumulative hazard rates or the survivor function leads to relatively smooth empirical curves, even when few observations are available. Graphs may yet be preferred, showing the hazard rates themselves, which provide points on which errors are independent. But in this case the curves are very chaotic and it is often difficult to evaluate their overall shape.

TABLE 7.1. Properties that make it possible to choose among several models

Function observed	Property	Model
$h(t)$	Independent of t	Exponential
$H(t)$	Linear function of t	Exponential
$\log h(t)$	Linear function of t	Gompertz
$\log (\log [\Delta S(t)])$	Linear function of t	Gompertz
$\log H(t)$	Linear function of $\log t$	Weibull
$\log (-\log S(t))$	Linear function of $\log t$	Weibull
$\log [\Delta S(t)]$	Linear function of t	Mover-stayer
$\log (\exp H(t) - 1)$	Linear function of $\log t$	Log-logistic
$\frac{1}{h(t)}$	Linear function of t	Pareto

Also, in many cases the small number of individuals observed make it possible to choose from among various possible distributions. In this case it is best to choose the distribution with the smallest number of parameters, but also the one that represents the most simple explicit form for the survivor function and hazard rates. As will be seen later, estimating the parameters is thus made easy when data including censored intervals are available. So a log-logistic distribution must be preferred to a log-normal distribution when the two fit the data to a same extent: as seen before, the log-logistic distribution features simple explicit forms for the survivor function and hazard rates.

7.2. Regression models

Parametric models, which have just been presented, imply working either on a homogeneous population, all individuals having the same probability of experiencing the event at a given time, or on a heterogeneous population, when this heterogeneity is not observed directly.

In fact, various characteristics of the respondents are often available, and we can assume that they influence the probability of experiencing an event. Thus, the educational level, the profession, and the social background of an individual must have an influence on his marriage, the birth of his children, the migrations he undertakes in the course of his life, etc. It is then of interest to generalize the previous models to take account of the heterogeneity observed between the individuals in the sample.

Various characteristics observed by the survey are thus available for a given individual. These can be represented in the form of a z vector:

$$z = (z_1, \ldots, z_s).$$

These characteristics can be quantitative variables (e.g. the number of migrations experienced by the individual in childhood, or his number of brothers and sisters) or qualitative variables, which are represented in a binary form (e.g. 0 if the individual is single, 1 if he is married at the beginning of the observation period).

The problem consists in modelizing the effect of the individual's various characteristics on the duration of his stay in the initial state.

It is possible to break down the observed population into various groups so that a non-parametric analysis can be carried out for each of these groups, and then to compare their behaviour patterns. Thus, one can compare the migration rates after age 15 for individuals who have experienced 0, 1, 2, 3, etc., migrations in childhood. In this case, the methods presented in previous chapters must be used.

Yet, when the number of groups increases, sub-populations soon reach figures that are too small to guarantee a reliable conclusion. Moreover, this method does not allow us to take account of characteristics that may change in the course of the individual's stay in the initial state. One can try and determine, for example, whether an individual's marriage decreases his probability to migrate.

Hence it is worthwhile using regression models which make it possible to take into account simultaneously the effect of the various characteristics. Of course, various hypotheses are necessary depending on which model is being used. It is useful to test these models before using them.

7.2.1. *Proportional hazard models*

The hypothesis underlying these models is that the various individual characteristics have a multiplicative effect on a hazard function which is the same for the whole population, over time. It follows that all individual hazard rates are proportional among themselves, whatever the time elapsed. If $h_0(t)$ stands for this underlying rate, which can be any one of the parametric forms presented before, the hazard rate of an individual with the z characteristics will take the form

$$h(t; z) = h_0(t) \exp(z\beta), \qquad (78)$$

with $z\beta = z_1\beta_1 + z_2\beta_2 + \ldots + z_n\beta_n$, where the column vector β represents the estimated effects of the various characteristics. It is clear that when all z variables are nil, we again have the basic model:

$$h(t; \mathbf{0}) = h_0(t). \qquad (79)$$

If only the z_1 variable is equal to the unit, all other variables being nil, it appears that

$$h(t; z_1) = h_0(t) \exp \beta_1, \qquad (80)$$

which leads to the relation

$$\frac{h(t; z_1)}{h(t; \mathbf{0})} = \exp \beta_1, \qquad (81)$$

which is independent of the form of the underlying rate $h_0(t)$. In fact, formally for two individuals n_i and n_j, the relation $h_{ni}(t)/h_{nj}(t)$ is a constant which depends on z_{ni} and z_{nj} but is independent of t. However, this ceases to be true when variables depending on time are introduced.

This relation of two conditional densities generalizes the epidemiologic concept of competing risks for two distinct groups. Although the parameter β_i measures the effect of variable z_i on the hazard rate, at times it is simpler to interpret $\exp(\beta_i)$ as a relative risk.[1]

Nevertheless, the concern that the data should verify this hypothesis of proportionality constitutes a source of discussion. Actually, these models are extremely general and non-restrictive (which is why they became popular). In cases where the hypothesis of proportionality is not entirely respected, the approximation they provide are often satisfactory. Otherwise, it is useful to break down the population into various sub-populations for which the characteristics that do not verify the hypothesis of proportionality are different (by sex, or by age at the beginning of the observation period).[2]

Violating the proportionality hypothesis implies that, in fact, there is an interaction between the duration (time) and one or several of the explanatory variables (which happens immediately if the variables depend on time).

Exponential distribution
In this case, the model is written

$$h(t; z) = \rho \exp(z\beta); \tag{82}$$

hence the density is

$$f(t; z) = \rho \exp(z\beta) \exp[-\rho t \exp(z\beta)] \tag{83}$$

and the survivor function is

$$S(t; z) = \exp[-\rho t \exp(z\beta)]. \tag{84}$$

---

[1] If z_i is dichotomic, it therefore separates individuals into two groups and $\exp(\beta_i)$ measures the relative risk of an individual in relation to those in the reference group. Otherwise, it can be said that, for an increase of the value of z_i of one unit, an individual's hazard rate is multiplied by $\exp(\beta_i)$.

[2] The non-parametric analysis previously carried out on the data generally provides quite an exhaustive overview of the behaviour patterns of the sub-populations to be distinguished.

Weibull distribution
In the case of a Weibull distribution, this model is written

$$h(t; z) = \lambda \rho (\rho t)^{\lambda-1} \exp(z\beta), \qquad (85)$$

leading to the survivor function,

$$S(t; z) = \exp[-(\rho t)^{\lambda} \exp(z\beta)] \qquad (86)$$

and the probability density,

$$f(t; z) = \lambda \rho (\rho t)^{\lambda-1} \exp[z\beta[(1-(\rho t)^{\lambda})]]. \qquad (87)$$

Gompertz distribution
In the case of a Gompertz distribution, the model is written

$$h(t; z) = \lambda \rho \exp(z\beta + \rho t), \qquad (88)$$

leading to the survivor function,

$$S(t; z) = \exp[\lambda(1 - \exp \rho t) \exp(z\beta)] \qquad (89)$$

and the probability density,

$$f(t; z) = \lambda \rho \exp[\rho t + z\beta + \lambda(1 - \exp \rho t)\exp z\beta]. \qquad (90)$$

d) Log-logistic distribution
The last example that will be given concerns the log-logistic distribution, which has the following hazard rate:

$$h(t; z) = \frac{\lambda \rho (\rho t)^{\lambda-1} \exp(z\beta)}{1 + (\rho t)^{\lambda}}, \qquad (91)$$

leading to the survivor function,

$$S(t; z) = [1 + (\rho t)^{\lambda}]^{-\exp z\beta} \qquad (92)$$

and the probability density,

$$f(t; z) = \frac{\lambda \rho (\rho t)^{\lambda-1} \exp z\beta}{[1 + (\rho t)^{\lambda}]^{1 + \exp z\beta}}. \qquad (93)$$

Checking the validity of the model
When characteristics are of a qualitative nature, each of them defines a sub-sample which possesses it (for example, the sub-sample of men compared with women, or the sub-sample of agricultural workers at the beginning of their stay). When they are of a quantitative nature,

Formalization of Parametric Analysis

then distinct sub-samples can again be defined (for example, the sub-sample of individuals with two brothers, or the sub-sample of individuals at the beginning of their stay between 20 and 24).

For all sub-populations in sufficient numbers, therefore, it is possible to estimate the hazard rates and to check whether the proportionality hypothesis still holds with regard to these results.

Thus, for instance, if the Weibull model is verified for the whole population, the values of $\log(-\log S(t; z_i))$ can be plotted against $\log t$. If the proportional hazard model is verified, this results in a series of lines that are parallel to one another.

More generally, indicating the values of $\log(-\log S(t; z_i))$ plotted against t results in a series of curves that are parallel to one another for the various characteristics, when the proportional hazard model is verified.

This constitutes, therefore, a means of testing the validity of the model. We have indicated in Fig. 7.16 the values of $\log(-\log S(t; z_i))$ plotted against $\log t$, with regard to the relationship between becoming a home-owner and being aged between 30 and 55, for women in the 'Triple Biographie' survey. The two curves correspond, the one to women with no diploma, the other to those with at least the 'Certificat d'études'. In this case the proportional hazard model is entirely valid, and a Weibull distribution fits the data between ages 30 and 45. On this point see the formula (54).)

When some characteristics do not have a multiplicative effect on instantaneous failure rates, it is possible to break down the population into several strata according to its characteristics, and to write, for instance, for strata j:

$$h_j(t; z) = h_{0j}(t) \exp(z\beta), \qquad (94)$$

where the parameters of the basic model can vary according to the strata considered. It is also possible to have models of different types according to the strata.

This model can be generalized by introducing a function $\psi(z; \beta)$ and by writing

$$h(t; z) = \psi(z; \beta)h_0(t). \qquad (95)$$

The function $\psi(z; \beta)$ must be equal to the unit for $z = 0$. It is possible to assume, for instance, that

$$h(t; z) = (1 + z\beta)h_0(t). \qquad (96)$$

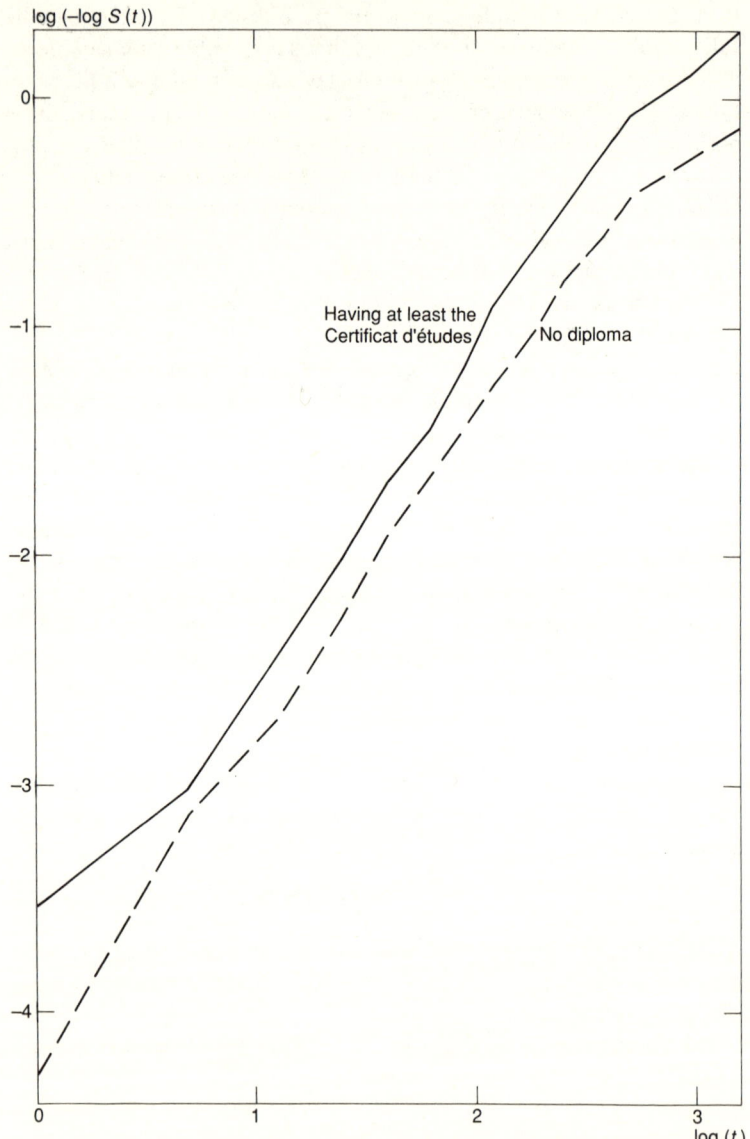

FIG. 7.16. Logarithm of the inverse of the logarithm of the survivor function plotted against t, for women who become home-owners between 30 and 35 years, according to whether they have no diploma or have obtained at least the Certificat d'études

Formalization of Parametric Analysis 141

In this case the effect of the variables is additive, the instantaneous failure rate being a linear function of these variables. Using this model can lead to an estimation of negative hazard rates, if we do not impose the condition

$$1 + z\beta > 0 \quad \text{for all } z.$$

Note that, as seen before, this does not happen when using a model of the multiplicative type, for which the hazard rate remains positive, whatever the values of the variables.

7.2.2. Accelerated failure time models

Let us now assume that the effect of the characteristics directly influences survivor functions, rather than instantaneous failure rates. In the multiplicative case, if the standard individual, for whom all variables are nil, has a survivor function equal to $S_0(t)$, then, for the individual with the z characteristics, we have the survivor function

$$S(t; z) = S_0(t \exp z\beta), \tag{97}$$

where the column vector β represents the estimated effects of the various characteristics. In that case, the probability density becomes

$$f(t; z) = f_0(t \exp z\beta) \exp z\beta, \tag{98}$$

and the hazard rate is written

$$h(t; z) = h_0(t \exp z\beta) \exp z\beta, \tag{99}$$

where $f_0(t)$ and $h_0(t)$ are the density and the hazard rate of an individual with all characteristics nil.

Therefore, we can write the model in terms of random variables:

$$T = T_0 \exp(-z\beta), \tag{100}$$

where T_0 is the duration of stay of an individual with all characteristics nil, who therefore has the survivor function $S_0(t)$. In that case, the various characteristics have a clear multiplicative effect on the duration of stay itself. Binary characteristics, for instance, speed up this duration of stay if their β_i is positive, or slow it down if it is negative. The model is therefore written

$$\log T = \log T_0 - z\beta. \tag{101}$$

Links between the accelerated failure time models and the proportional hazard model

We shall show that, in fact, some models feature the two previous properties. They must verify the following relation:

$$h_0^1(t) \exp(z\beta_1) = h_0^2(t \exp z\beta_2) \exp(z\beta_2), \qquad (102)$$

where indexes 1 and 2 stand for the proportional hazard model and the accelerated failure time loss model. This relation must be verified for all values of t and z. For $z = 0$, it is written

$$h_0^1(t) = h_0^2(t) = h_0(t), \qquad (103)$$

a relation that is valid for all values of t. By summing on all values of t, we obtain

$$H_0(t) \exp(z\beta_1) = H_0(t \exp(z\beta_2)). \qquad (104)$$

It is clear that the only function $H_0(t)$ that makes it possible to verify this relation is the function

$$H_0(t) = (\rho t)^\lambda, \qquad (105)$$

and vector β_1 must be proportional to vector β_2, with the proportionality coefficient being λ. In fact, the previous relation (104) is written by taking its logarithm:

$$\lambda \log(\rho t) + z\beta_1 = \lambda \log(\rho t \exp z\beta_2) = \lambda \log(\rho t) + \lambda z\beta_2. \qquad (106)$$

It appears that the Weibull distribution or exponential distribution, when $\lambda = 1$, are the only distributions that verify both the proportional hazard and the accelerated failure time models.

Log-logistic distribution

Among the accelerated failure time models, it is interesting to consider the one that corresponds to the log-logistic distribution. If

$$S_0(t) = (1 + (\rho t)^\lambda)^{-1}, \qquad (107)$$

it follows that

$$S(t; z) = [1 + (\rho t \exp z\beta)^\lambda]^{-1}. \qquad (108)$$

Hence

$$f(t; z) = \frac{\lambda \rho (\rho t \exp z\beta)^{\lambda-1} \exp z\beta}{[1 + (\rho t \exp z\beta)^\lambda]^2}. \qquad (109)$$

and

$$h(t;z) = \frac{\lambda \rho (\rho t \exp z\beta)^{\lambda-1} \exp z\beta}{1 + (\rho t \exp z\beta)^{\lambda}}. \tag{110}$$

In that case, the model is different from the log-logistic proportional hazard model.

The hazard rate is written in a simpler way by introducing a first parameter β_1, which verifies

$$\exp(\beta_1) = \rho^{\lambda},$$

and by substituting β for $\lambda\beta$:

$$h(t;z) = \frac{\lambda}{t\,[1 + t^{-\lambda} \exp(-z\beta)]}. \tag{111}$$

Checking the validity of the model

As before, the total sample can be broken down into sub-populations of sufficient size, corresponding to the various characteristics considered. In this case, the formula (101) indicates that the distributions of log T for various values of z can be deducted from one another by translation. It follows, for instance, that the variance of log T is independent of the characteristic that defines the sub-population. It is then sufficient to compute this variance for various sub-populations and to check that it is approximately constant. In order to do that, however, there should not be any censored intervals. Because, generally, there are censored intervals, it is necessary to consider each type of distribution separately in order to determine the tests that can be carried out.

Thus, for instance, with a log-logistic distribution, we can compute

$$\log\left(\frac{1}{S(t)} - 1\right) = \lambda \log \rho t + \lambda z \beta. \tag{112}$$

In this case, if the model is of an accelerated failure time model, $\log[(1/S(t)) - 1]$ is a linear function of log t, and for the various values of the characteristics, each of these lines always has the same slope: λ.

7.2.3. *More complex models*

Analysing the parameters of a model can result in making them depend, in part or entirely, on the various characteristics. Here we briefly present the generalized Gompertz–Makeham model which can be estimated using the RATE program (cf. Appendix). This model can be written

$$h(t) = \lambda_1 \exp(z_1\beta_1) + \lambda\rho \exp[z_2\beta_2 + \rho(1+z_3\beta_3)t]. \qquad (113)$$

It appears that the hazard rate depends on three series of variables z_1, z_2, and z_3, which can be different, but can also contain common variables. This program also allows a substitution of $(1+z\beta)$ for $\exp(z\beta)$ and reciprocally. Again, there is a risk that the linear model might lead to negative estimated hazard rates for some values of the characteristics.

7.3. Conclusion

This chapter has allowed us to show the large variety of the parametric functions that can be chosen to represent various types of distribution. In practice, it is often difficult to choose from among distributions that are close to one another. In this case one must choose the distribution for which there is a simple and functional form of the hazard rates and survivor function. The reasons for this choice will appear more clearly in the next chapter, which deals with methods of estimating of the parameters of these functions.

It has also enabled us to take account of various characteristics of the individuals observed and to modelize their effect on hazard rates, the survivor function, or the probability density. Again, it is better to use rather simple models, with additive or multiplicative effects, in order to be able to estimate the parameters that play a part in these models. Before applying a given type of model, however, it is necessary to check that it correctly fits the characteristics studied.

8

Methods of Estimation of Parametric Models

In the previous chapter, we discussed various parametric models and showed the preliminary analyses that make it possible to choose a model adapted to the data. In this respect, it is very useful to break down the whole population into sub-populations for whom the various characteristics do or do not apply.

Let us now suppose that one of these models has been chosen. It is then necessary to estimate the various parameters of the model, as well as the coefficients to be applied to the characteristics, in order to represent the full data-set with greater accuracy. An estimate of the variance of these parameters or estimated coefficients should also be available, as well as their covariances between these parameters, in order to be able to carry out a certain number of tests. It is useful, for example, in the case of a multiplicative model, to check whether the effect of a characteristic is markedly different from the unit—that is, whether or not this characteristic influences the phenomena being studied. It is also useful to compare the effects of several characteristics. When identical, these effects can be replaced by a single characteristic. Thus, for instance, if the fact of being married, widowed, or divorced influences the probability to migrate in the same way, these three characteristics can be replaced by a single one—the fact of not being single any longer.

Data from surveys are generally censored, which must be taken into account in the methods of estimation. The various types of censoring that may occur in surveys have already been indicated.

We shall first explain how to calculate the likelihood of an observation, according to whether it is censored or not. Then, we shall present the general methods of estimation of the various parameters or coefficients, when the samples available are large enough. These methods will be developed in various particular cases, giving precise examples as an illustration. Lastly, we shall examine in further detail the problems raised by the choice between different parametric models, which will lead us to the following chapter on semi-parametric models.

8.1. Computation of the Likelihood in the Presence of Censoring

The maximum likelihood estimator has already been introduced in the non-parametric case (Chapter 2). Here we address it in the context of parametric models.

The survey observes n individuals and collects information on the events that have occurred over a certain lapse of time. Let us first consider the case where that interval includes the date of entry into the population at risk, but may not include the date at which the individual actually experiences the event under study. In this case, right-censored intervals are observed. These intervals, however, do provide some information, for we know that before a given date the individual has not yet undergone the event. Let us represent all the information gathered in the form of a triplet:

$$(t_i^0, \delta_i, z_i) \quad \text{with} \quad i = 1, \ldots, n,$$

where t_i^0 is the duration of stay in the initial state, either up to the occurrence, if the event is indeed experienced ($\delta_i = 1$), or up to the date when the individual ceases to be observed before the occurrence of the event ($\delta_i = 0$), and where z_i is the vector of the characteristics of the individual at the beginning of his stay. Therefore, δ_i is a binary variable indicating whether or not the event has occurred before the individual is lost.

This censoring can take various forms. Generally, there is absolutely no relation between the censoring time and the fact that the individual has or has not experienced the event. In such a case, we have *random losses* with regard to the event studied. It is then possible to introduce a random variable T^s, with survivor function $O_i(t)$ and probability density $q_i(t)$. This variable is independent of the failure time of T, which has a survivor function $S(t; z_i, \beta)$ and a probability density $f(t; z_i, \beta)$, where β stands for all the parameters and coefficients to be estimated. The variable that is in fact observed is written

$$T^0 = \min(T^s, T). \tag{1}$$

These notations make it possible to calculate the probability

$$P(t \leq T_i^0 < t + dt, \delta_i = 1; z_i, \beta) = O_i(t) f(t; z_i, \beta) dt. \qquad (2)$$

In fact, in this case the individual has experienced the event at time t, observed ($\delta_i = 1$), and therefore is not yet censored at that time. As the two events are independent, it is indeed the product of the probability that the individual is not lost at time t, ($O_i(t)$) times the probability that he has experienced the event at the same time ($f(t; z_i, \beta)$) that should be calculated. Similarly, for the respondents who have not yet experienced the event, we can write

$$P(t \leq T_i^0 < t + dt, \delta_i = 0; z_i, \beta) = q_i(t) S(t; z_i, \beta) dt. \qquad (3)$$

As neither $O_i(t)$ nor $q_i(t)$ provide any information on the parameters and coefficients to be estimated, β, the likelihood of the data, can then be considered as proportional to the quantity

$$L(\beta) = \prod_{i=1}^{n} f(t_i; z_i, \beta)^{\delta_i} S(t_i; z_i, \beta)^{1-\delta_i} = \prod_{i=1}^{n} h(t_i; z_i, \beta)^{\delta_i} S(t_i; z_i, \beta), \qquad (4)$$

where $h(t; z, \beta)$ is the instantaneous failure rate.

In this case, knowing the distribution of hazard rates and of survivor functions makes it possible to estimate the values of the various parameters by the maximum likelihood method. This is considerably easier when these two distributions have a simple explicit form. On the other hand, when one of them takes a more complex form, calculations are very complex too, and can even become entirely impracticable.

This likelihood remains valid in other situations where censoring is no longer random. In particular, it can still be applied when losses occur at each time independently of the risks to which the individuals are subjected. In this case, the rules of these losses can depend on the previous history of the event under study, but must not eliminate preferentially individuals whose risk of experiencing the event is high or, conversely, low.

It is then said that there is a system of *independent censoring*.[1] This is the case when observation is stopped after the occurrence of the r th event, for instance, or when we remove from the observation a given fraction of the individuals at risk after the occurrence of events of a given rank.

---

[1] For a demonstration of the validity of the likelihood, see Kalbfleish and Prentice (1980: 119–22).

8.2. Estimating the Parameters and Testing their Value

Let us rewrite the likelihood of the observations, given in the previous section, in a logarithmic form:

$$\log L(\beta) = \sum_{i=1}^{n} [\delta_i \log h(t_i; z_i, \beta) + \log S(t_i; z_i, \beta)]. \tag{5}$$

The method consists of giving the β parameters the values that maximize this likelihood. To that end, we can work on the first derivative of $\log L(\beta)$ in relation to β. The solutions of the equation equalling this to zero provide these values.

When the number of observations is large enough, it can be shown that under simple conditions, generally met (i.e., $L(\beta)$ must be thrice differentiable, and certain boundedness conditions on this third derivative must be satisfied[2]), this equation has a single solution. The estimation of the parameters thus obtained has an asymptotical mean equal to 0 with a minimum variance. The asymptotic distribution of this estimator is multivariate normal with as many parameters as there are parameters to be estimated. The mean of this distribution is the 'true' value of β. If we calculate the inverse of the matrix of the second derivative of $\log L(\beta)$, also called the Fisher information matrix, we show that the inverse of that matrix represents an estimation of the matrix of variances and covariances of the parameters β (see Section 3.2.1).

It is then possible to carry out various tests on the estimated β parameters. Thus, for example, we calculate the quantity $(\hat{\beta}_i^2 / \text{var } \hat{\beta}_i)$ to determine whether a parameter β_i can be considered as significantly different from zero. If it is not, this statistic is a χ^2 with one degree of freedom. Similarly, it is possible to test, more generally, whether the estimated β parameters are different from the values β_0 which were chosen *a priori*. If the matrix of variances and covariances is denoted $V(\beta)$, it is still possible to write the statistic

$$(\hat{\beta} - \beta_0)^T V(\hat{\beta})^{-1} (\hat{\beta} - \beta_0), \tag{6}$$

which, when parameters β are not significantly different from β_0, is a χ^2 with as many degrees of freedom as there are parameters considered.

---

[2] On this subject, see Cox and Hinkley (1974: 281 *et seq.*).

Estimation of Parametric Models

It is also possible to use the likelihood directly by computing the ratio

$$R(\beta_0) = \frac{L(\beta_0)}{L(\hat{\beta})}. \tag{7}$$

To test the hypothesis that the estimated β parameters are not different from the given β_0 values, the asymptotic distribution of $(-2 \log R(\beta_0))$ is a χ^2 with as many degrees of freedom as there are parameters to be considered. This method can easily be generalized, to test the value of any number of parameters.

Lastly, it is possible to use the first derivative of the likelihood

$$U(\beta) = \frac{\partial \log L(\beta)}{\partial \beta}. \tag{8}$$

When $\beta = \beta_0$, the statistic is asymptotically normal, of mean 0 and variance $V(\beta_0)$. Under these conditions, the statistic

$$U'(\beta_0) \, V(\beta_0)^{-1} \, U(\beta_0) \tag{9}$$

has an asymptotic distribution of χ^2 with as many degrees of freedom as there are parameters to be considered. It is then possible to determine whether the proposed value β_0 of the parameters can be considered as satisfactory, given the observations.

We shall not proceed further with this likelihood theory. Further details on these methods can be found in the work of Cox and Hinkley (1974). We shall now examine precise cases of the estimation.

8.3. Estimation of the Parameters: Examples

To show more clearly how this estimation can be carried out, we shall consider different models and give concrete examples.

8.3.1. Exponential model

The exponential model represents the simplest case, with no explanatory variables. Suppose we observe n individuals. Let d be the number of occurrences at times t_i, $i = 1, \ldots, d$. Each individual experiences one event at the most. In that case, $(n - d)$ censored intervals are observed, with the observation ending at times t_j, $j = (d + 1), \ldots, n$.

If the hazard rate is equal to ρ, which is the only parameter to be estimated, then applying formula (5) leads to the following log-likelihood

$$\log L(\rho) = \sum_{i=1}^{d} \log \rho - \sum_{i=1}^{n} \rho t_i = d \log \rho - \rho \sum_{i=1}^{n} t_i. \qquad (10)$$

It can then be said that $\sum_{i=1}^{n} t_i$ is the total time spent at risk by the individuals. We can then write

$$U(\rho) = \frac{\partial \log L(\rho)}{\partial \rho} = \frac{d}{\rho} - \sum_{i=1}^{n} t_i \qquad (11)$$

and

$$V(\rho)^{-1} = -\frac{\partial^2 \log L(\rho)}{\partial \rho^2} = \frac{d}{\rho^2}. \qquad (12)$$

The maximum likelihood estimator $\hat{\rho}$ of ρ is the solution to the equation

$$U(\rho) = \frac{d}{\rho} - \sum_{i=1}^{n} t_i = 0, \qquad (13)$$

which gives

$$\hat{\rho} = \frac{d}{\sum_{i=1}^{n} t_i} \qquad (14)$$

Thus, it is the ratio of the number of events to the total time spent at risk by the individuals.

Relation (12) gives us the estimated variance of ρ:

$$V(\hat{\rho}) = \frac{d}{\left(\sum_{i=1}^{n} t_i\right)^2}. \qquad (15)$$

It is thus possible to build the following 95 per cent confidence interval for ρ:

$$\frac{d - 1.96\sqrt{d}}{\sum_{i=1}^{n} t_i} < \rho < \frac{d + 1.96\sqrt{d}}{\sum_{i=1}^{n} t_i}. \qquad (16)$$

Estimation of Parametric Models 151

It is also possible to use the logarithm of the likelihood ratio:

$$-2 \log R(\rho) = 2 \left[\rho \sum_{i=1}^{n} t_i - d \log \rho + d \left(\log \frac{d}{\sum_{i=1}^{n} t_i} - 1 \right) \right], \quad (17)$$

which has a χ^2 asymptotic distribution with one degree of freedom. Thus, there is a 95 per cent confidence interval, for instance, for those values of ρ for which the function has a value inferior to 3.84.

Lastly, it is possible to use the first derivative of the likelihood, resulting in the statistic

$$\left(\frac{d}{\rho} - \sum_{i=1}^{n} t_i \right)^2 \frac{\rho^2}{d} = d + \frac{\rho^2}{d} \left(\sum_{i=1}^{n} t_i \right)^2 - 2\rho \sum_{i=1}^{n} t_i, \quad (18)$$

which has a χ^2 asymptotic distribution with one degree of freedom. There is again a 95 per cent confidence interval when this function takes a value inferior to 3.84. It follows that the square root of the statistic

$$\left(\frac{d}{\rho} - \sum_{i=1}^{n} t_i \right) \frac{\rho}{\sqrt{d}} = \sqrt{d} - \frac{\rho}{\sqrt{d}} \sum_{i=1}^{n} t_i \quad (19)$$

has a normal distribution, resulting in a 95 per cent confidence interval for ρ, which is identical to the one given by formula (16). Though different in theory, the two methods lead to an identical procedure.

Let us apply these methods to the probability of becoming a homeowner after the birth of the last child for women whose spouses are skilled workers. These data are still taken from the 'Triple Biographie' survey. Of the 380 women, 137 become home-owners before the survey, which gives $d = 137$. These 380 women have spent 8307 years without owning their dwelling, which gives

$$\sum_{i=1}^{n} t_i = 8307.$$

With these figures, it is possible to obtain the estimator of the parameter ρ, using the maximum likelihood method:

$$\hat{\rho} = \frac{137}{8307} = 0.0165.$$

The variance and the standard deviation of this estimation are

$$V(\hat{\rho}) = \frac{137}{(8307)^2} = 1.9853 \times 10^{-6} \quad \text{and} \quad \sigma(\hat{\rho}) = 1.409 \times 10^{-3}$$

This leads to the following 95 per cent confidence interval for ρ, when $\sqrt{d} = 11.705$:

$$\frac{137 - 1.96 \times 11.705}{8307} < \rho < \frac{137 + 1.96 \times 11.705}{8307},$$

which gives $0.0137 < \rho < 0.0192$. The maximum of the logarithm of the likelihood is equal to

$$\log L(\hat{\rho}) = -699.3346.$$

This makes it possible to draw Fig. 8.1, which reports the logarithm of that likelihood around ρ, of which the equation is

$$-2 \log R(\rho) = 2(8307\,\rho - 137 \log \rho - 699.3346).$$

The 95 per cent confidence interval is obtained by drawing a parallel to the abscissa axis, at the ordinate point 3.481, a value for which there are only 5 chances out of 100 for ρ to be outside this interval. This results in the confidence interval

$$0.0139 < \rho < 0.0194.$$

This interval is slightly different from the previous one, because the curve is not symmetrical, but the results obtained through the various methods are consistent.

Let us now introduce the effect of some characteristics, in the form of a regression vector. The instantaneous failure rate for individual i can be written

$$h(t; z_i) = \exp(z_i \beta). \tag{20}$$

To each individual corresponds the vector $z_i = (z_{i1}, \ldots, z_{is})$, in which the first element is equal to the unit, so that $\rho = \exp(\beta_1)$, that is, the hazard rate that corresponds to the individuals whose other characteristics are all nil. Then we can write

$$\log L(\beta) = \sum_{i=1}^{n} z_i \beta - \sum_{i=1}^{n} t_i \exp(z_i \beta). \tag{21}$$

For the j th characteristic, it follows that

$$U_j(\beta) = \frac{\partial \log L(\beta)}{\partial \beta_j} = \sum_{i=1}^{d} z_{ij} - \sum_{i=1}^{n} z_{ij}\, t_i \exp(z_i \beta), \qquad (22)$$

where z_{ij} is the j th characteristic of the individual i. Similarly, as an element of the j th line and of the k th column, the opposite of the matrix of the second derivative contains

$$-\frac{\partial^2 \log L(\beta)}{\partial \beta_j\, \partial \beta_k} = \sum_{i=1}^{n} z_{ij}\, z_{ik}\, t_i \exp(z_i \beta). \qquad (23)$$

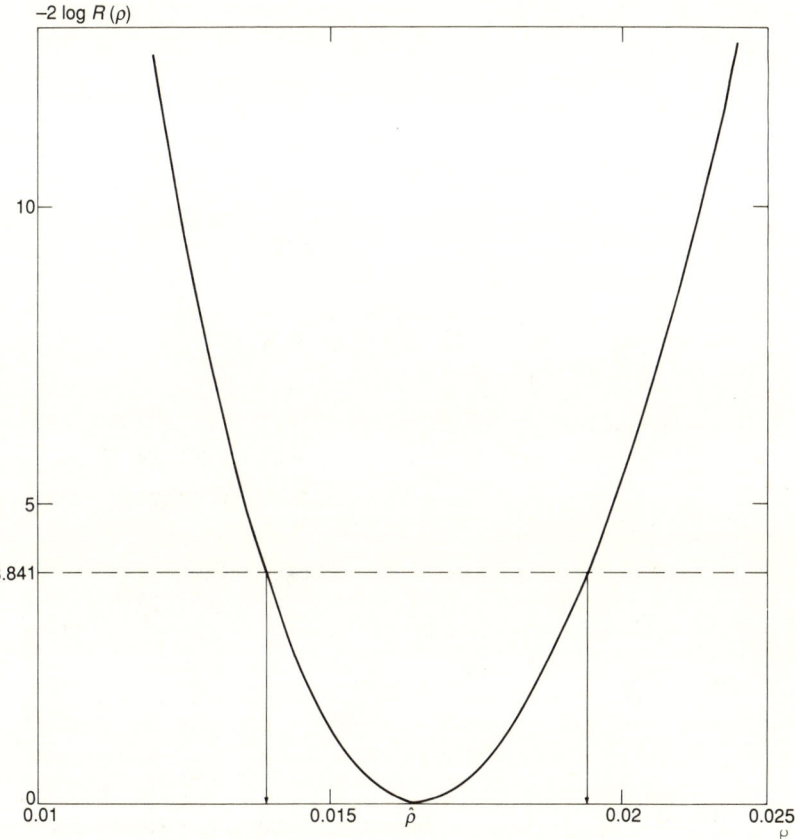

FIG. 8.1. Logarithm of the likelihood in relation to parameter ρ, and computation of the confidence interval around the estimation $\hat{\rho}$, for the probability of skilled workers becoming home-owners after the birth of their last child

The maximum likelihood method implies adopting the solution of the system of equations:

$$\sum_{i=1}^{d} z_{ij} - \sum_{i=1}^{n} z_{ij} t_i \exp(z_i \beta) = 0, \quad j = 1, \ldots, n. \quad (24)$$

The next sub-section considers a method for solving this system in the general case. Here, let us consider the simple case, taking account of only one binary variable. In that case, the vector $z_i = (1, z_{i2})$ and the log-likelihood is written

$$\log L(\beta) = \sum_{i=1}^{d} (\beta_1 + z_{i2} \beta_2) - \sum_{i=1}^{n} t_i \exp(\beta_1 + z_{i2} \beta_2), \quad (25)$$

resulting in the two derivatives,

$$U_1(\beta) = \frac{\partial \log L(\beta)}{\partial \beta_1} = d - \sum_{i=1}^{n} t_i \exp(\beta_1 + z_{i2} \beta_2) = 0$$

$$U_2(\beta) = \frac{\partial \log L(\beta)}{\partial \beta_2} = \sum_{i=1}^{d} z_{i2} - \sum_{i=1}^{n} t_i z_{i2} \exp(\beta_1 + z_{i2} \beta_2) = 0. \quad (26)$$

This system can be written in a simpler way by taking into account the number of events observed when $z_{i2} = 0(d_0)$, the number of events observed when $z_{i2} = 1(d_1)$, and with N_0 and N_1 the total time spent at risk by the individuals in the two categories:

$$U_1(\beta) = d_0 + d_1 - N_0 \exp \beta_1 - N_1 \exp(\beta_1 + \beta_2) = 0$$

$$U_2(\beta) = d_1 - N_1 \exp(\beta_1 + \beta_2) = 0. \quad (27)$$

It follows that

$$\hat{\beta}_1 = \log \frac{d_0}{N_0} \quad \hat{\beta}_1 = \log \frac{d_1 N_0}{d_0 N_1} \quad (28)$$

In that case, the opposite of the matrix of the second derivative is equal to

$$\begin{pmatrix} N_0 \exp \beta_1 + N_1 \exp(\beta_1 + \beta_2) & N_1 \exp(\beta_1 + \beta_2) \\ N_1 \exp(\beta_1 + \beta_2) & N_1 \exp(\beta_1 + \beta_2) \end{pmatrix} = \begin{pmatrix} d_0 + d_1 & d_1 \\ d_1 & d_1 \end{pmatrix}, \quad (29)$$

Estimation of Parametric Models

resulting in the matrix of variances and covariances which is independent of N_1 and N_2:

$$\frac{1}{d_0 d_1} \begin{pmatrix} d_1 & -d_1 \\ -d_1 & d_0 + d_1 \end{pmatrix}. \tag{30}$$

Thus, if the variance of β_1 is d_0^{-1}, that of β_2 is $(d_0^{-1} + d_1^{-1})$, and the covariance of these two parameters is negative and equal to $-d_0^{-1}$. It is therefore possible to test an *a priori* value of the two parameters using the previously indicated methods.

These results can easily be generalized when variable z_2 takes integer values equal to 0, 1, 2, etc.

As an illustration, let us consider again the probability of becoming a home-owner after the birth of the last child for women whose spouses are skilled workers. We want to take into account the educational level of these women and determine whether it influences that probability. As a first solution, we can break down the sample into sub-samples of given educational levels and estimate, as before, their parameter ρ. The women have been distinguished according to whether their spouse had no diploma (0), the Certificat d'études (1), or the Certificat d'aptitude professionnelle or Brevet d'études primaires (2). Only three skilled workers had a higher level diploma: we left them out.

The logarithm of the parameter is estimated for the three sub-populations:

$$\log \hat{\rho}_0 = -4.4257$$

$$\log \hat{\rho}_1 = -3.8863$$

$$\log \hat{\rho}_2 = -3.3779.$$

It appears that, between the first and second population, 0.5394 must be added to the logarithm of $\hat{\rho}_0$, and between the second and third, 0.5084 must be added to the logarithm of $\hat{\rho}_1$. It is then possible to use a synthetic variable, taking values 0, 1, or 2 according to the individual's diplomas.

Let d_0, d_1, d_2 be the numbers of women married to skilled workers with diplomas 0, 1, 2 who became home-owners before the survey. If all these women have spent N_0, N_1, N_2 years without owning their dwellings, we show, as before, that the derivatives of the log-likelihood in relation to β_1 and β_2 are

$$U_1(\beta) = d_0 + d_1 + d_2 - N_0 \exp \beta_1 - N_1 \exp (\beta_1 + \beta_2)$$
$$- N_2 \exp (\beta_1 + 2\beta_2) = 0$$
$$U_2(\beta) = d_1 + 2d_2 - N_1 \exp (\beta_1 + \beta_2)$$
$$- 2N_2 \exp (\beta_1 + 2\beta_2) = 0 \quad (31)$$

In the present case,

$$d_0 = 54, \quad d_1 = 70, \quad d_2 = 13$$
$$N = 4513, \quad N_1 = 3412, \quad N_2 = 381.$$

The two-equation system (31) is then easily solved and results in the estimations

$$\hat{\beta}_1 = -4.422$$
$$\hat{\beta}_2 = 0.5299.$$

The negative of the matrix of the second derivative is therefore written

$$\begin{pmatrix} N_0 \exp \beta_1 + N_1 \exp (\beta_1 + \beta_2) + N_2 \exp (\beta_1 + 2\beta_2) & N_1 \exp (\beta_1 + \beta_2) + 2N_2 \exp (\beta_1 + 2\beta_2) \\ N_1 \exp (\beta_1 + \beta_2) + 2N_2 \exp (\beta_1 + 2\beta_2) & N_1 \exp (\beta_1 + \beta_2) + 4N_2 \exp (\beta_1 + \beta_2) \end{pmatrix}, \quad (32)$$

which is equal to

$$\begin{pmatrix} d_0 + d_1 + d_2 & d_1 + 2d_2 \\ d_1 + 2d_2 & d_1 + 2d_2 + 2N_2 \exp (\beta_1 + 2\beta_2) \end{pmatrix}. \quad (33)$$

This provides the matrix of variances and covariances, which depends on values of N:

$$\frac{1}{(d_0 + d_1 + d_2)[d_1 + 2d_2 + 2N_2 \exp (\beta_1 + 2\beta_2)] - (d_1 + 2d_2)^2}$$
$$\times \begin{pmatrix} d_1 + 2d_2 + 2N_2 \exp (\beta_1 + 2\beta_2) & -(d_1 + 2d_2) \\ -(d_1 + 2d_2) & d_0 + d_1 + d_2 \end{pmatrix} \quad (34)$$

Here, this matrix is written

$$\begin{pmatrix} 0.0162 & -0.0127 \\ -0.0127 & 0.0181 \end{pmatrix}.$$

Estimation of Parametric Models 157

It is then possible to determine whether the educational level influences in a significant way the probability of becoming a home-owner, by computing the χ^2 statistic with one degree of freedom:

$$\frac{\hat{\beta}_2^2}{V(\hat{\beta}_2)} = \frac{0.281}{0.0181} = 15.52.$$

The effect is really significant: the higher the educational level, the greater the probability of becoming a home-owner.

8.3.2. *The Newton–Raphson technique*

In the case of an exponential model, as soon as the number of parameters to estimate exceeds two, solving the equations systems $U_j(\beta) = 0$ quickly becomes very complex. In this case, an approach through successive iterations can be useful, although it can sometimes lead to improper solutions, as we shall show further on.

The Newton–Raphson technique is thus regularly used. Let us consider the likelihood of the observations, $L(\beta)$, which depends on β parameters. These parameters are estimated like the parameters that maximize the likelihood and therefore cancel out its derivative, also called the score statistic. The Newton–Raphson technique is therefore based on the first-order Taylor series expansion of

$$U(\beta) = \frac{d \log L(\beta)}{d \beta}.$$

Given a β_0 value, the first-order expansion is

$$U(\hat{\beta}) = U(\beta_0) - I(\beta^*)(\hat{\beta} - \beta_0), \tag{35}$$

where β^* is a value 'between' $\hat{\beta}$ and β_0, and where $I(\beta) = -d^2 \log L(\beta)/d\beta^2$, the Fisher information matrix (see Section 3.2.1), of which the opposite is the estimator of variances and covariances for the coordinates of vector $\hat{\beta}$. If β_0 is close to $\hat{\beta}$, then $I(\beta^*) \simeq I(\beta_0)$. Knowing that $U(\hat{\beta}) = 0$, formula (35) becomes

$$\hat{\beta} = \beta_0 + I(\beta_0)^{-1} U(\beta_0). \tag{36}$$

The right-hand member gives a new trial value of β. The whole process is repeated until the estimation of β, $\hat{\beta}$ converges towards an acceptable solution of $U(\hat{\beta}) = 0$.

Generally, this method gives accurate results when the likelihood is unimodal, even if the initial value β_0 is far from the $\hat{\beta}$ value to be estimated. On the other hand, if the distribution is multimodal, the method can lead to a relative maximum that is different from the real one and thus can provide totally inaccurate results. To assess whether the solution reached is the correct one, it is useful to work with various initial β_0 values and to check whether the same estimation is always obtained for $\hat{\beta}$. If not, the one with the maximum likelihood should be chosen.

8.3.3. The Weibull model

Let us first consider the Weibull model without taking account of explanatory variables.

With d events always observed on n individuals, who cease to be observed (event or loss) on times t_i, the log-likelihood is written

$$\log L(\lambda, \rho) = \sum_{i=1}^{d} [\log \lambda + \log \rho + (\lambda - 1) \log \rho t_i] - \sum_{i=1}^{n} (\rho t_i)^{\lambda}. \quad (37)$$

The derivative in relation to the two parameters of this function are written

$$\frac{\partial \log L(\lambda, \rho)}{\partial \lambda} = \frac{d}{\lambda} + d \log \rho + \sum_{i=1}^{d} \log t_i - \rho^{\lambda} \sum_{i=1}^{n} t_i^{\lambda} \log(\rho t_i) = 0.$$

$$\frac{\partial \log L(\lambda, \rho)}{\partial \lambda} = \frac{\lambda d}{\rho} - \lambda \rho^{\lambda - 1} \sum_{i=1}^{n} t_i^{\lambda} = 0. \quad (38)$$

The first equation gives ρ in relation to λ:

$$\rho = \left(\frac{d}{\sum_{i=1}^{n} t_i^{\lambda}} \right)^{1/\lambda}. \quad (39)$$

Transferring this value of ρ into the second equation gives

$$f(\lambda) = \frac{d}{\lambda} + \sum_{i=1}^{d} \log t_i - d \frac{\sum_{i=1}^{n} t_i^{\lambda} \log(t_i)}{\sum_{i=1}^{n} t_i^{\lambda}} = 0. \quad (40)$$

Estimation of Parametric Models

This equation can be solved through successive approximations using its derivative in relation to λ:

$$f'(\lambda) = -\frac{d}{\lambda^2} - d \frac{\left(\sum_{i=1}^{n} t_i^\lambda (\log t_i)^2\right)\left(\sum_{i=1}^{n} t_i^\lambda\right) - \left(\sum_{i=1}^{n} t_i^\lambda \log t_i\right)^2}{\left(\sum_{i=1}^{n} t_i^\lambda\right)^2} \quad (41)$$

From a λ_0 value, we estimate a new λ_1 corrected value of the relation $f(\lambda_0)/f'(\lambda_0)$. This is done until the $f(\lambda_0)$ value obtained is close enough to zero, at a threshold that has been fixed in advance. We thus obtain $\hat{\lambda}$, which, once included in formula (39), also gives $\hat{\rho}$.

Consider now the second derivative of $l = \log L(\lambda, \rho)$:

$$-\frac{\partial^2 l}{\partial \lambda^2} = \frac{d}{\lambda^2} + \rho^\lambda \sum_{i=1}^{n} t_i^\lambda [\log(\rho t_i)]^2$$

$$-\frac{\partial^2 l}{\partial \rho^2} = \frac{\lambda d}{\rho^2} + \lambda(\lambda - 1) \rho^{\lambda-2} \sum_{i=1}^{n} t_i^\lambda \quad (42)$$

$$-\frac{\partial^2 l}{\partial \lambda \, \partial \rho} = -\frac{d}{\rho} + \rho^{\lambda-1}(1 + \lambda \log \rho) \sum_{i=1}^{n} t_i^\lambda + \lambda \rho^{\lambda-1} \sum_{i=1}^{n} t_i^\lambda \log t_i.$$

$\hat{\lambda}$ and $\hat{\rho}$ being available, we can calculate the matrix of variances and covariances of these parameters using these three formulas and test various supposed values for these parameters.

Let us apply this same model to the probability of becoming a home-owner after the birth of the last child for women married to workers. Estimating the parameters by the method presented here leads to

$$\hat{\lambda} = 1.0200$$

$$\hat{\rho} = 0.0168.$$

The $\hat{\rho}$ parameter estimated using this model is very close to the parameter $\hat{\rho}$ estimated with the exponential model ($\hat{\rho} = 0.0165$), and the value of parameter $\hat{\lambda}$ is close to the unit. These results are consistent with the fact that the Weibull model becomes identical to the exponential model when $\lambda = 1$.

It is then useful to determine whether or not $\hat{\lambda}$ can be considered as different from the unit. To do so, one can use the variances estimated with the equations (42):

$$V(\hat{\lambda}) = 6.29840 \times 10^{-3}$$

$$V(\hat{\rho}) = 3.22399 \times 10^{-6}$$

$$\text{cov}(\hat{\lambda}\,\hat{\rho}) = 8.86654 \times 10^{-5}.$$

If the equality hypothesis is verified, variable $(\hat{\lambda} - 1)$ is normal with mean 0 and standard deviation $\sigma(\hat{\lambda}) = 7.936 \times 10^{-2}$. The value 0.02 perfectly verifies the equality hypothesis. The exponential model fits the data.

One can also calculate $f(1)$, first derivative of the log-likelihood for $\lambda = 1$ and $\hat{\rho}_1 = d/\sum_{i=1}^{n} t_i = 0.0165$:

$$f(1) = d + \sum_{i=1}^{d} \log t_i - d \frac{\sum_{i=1}^{n} t_i \log t_i}{\sum_{i=1}^{n} t_i} = 3.2326. \quad (43)$$

The opposed of the matrix of the second derivative at point $(1, \hat{\rho}_1)$ is simply written in the form

$$-\frac{\partial^2 l}{\partial \lambda^2} = d + \hat{\rho}_1 \sum_{i=1}^{n} t_i [\log(\hat{\rho}_1 t_i)]^2$$

$$-\frac{\partial^2 l}{\partial \rho^2} = \frac{d}{\hat{\rho}_1^2} \quad (44)$$

$$-\frac{\partial^2 l}{\partial \lambda \, \partial \rho} = \sum_{i=1}^{n} t_i \log(\hat{\rho}_1 t_i).$$

It is then possible to estimate the terms of the opposite matrix, which in this case is written

$$\begin{pmatrix} 6.0522 \times 10^{-3} & 8.791 \times 10^{-5} \\ 8.791 \times 10^{-5} & 3.258 \times 10^{-6} \end{pmatrix}.$$

If the hypothesis is verified, the quantity

$$3.2326 \times \sqrt{6.0522 \cdot 10^{-3}} = 0.252$$

Estimation of Parametric Models

can also be correctly considered as drawn according to a standard normal trial.

Let us now introduce the characteristics z of the various individuals. In the multiplicative case, the Weibull model is written

$$h(t, z) = \lambda \rho (\rho t)^{\lambda - 1} \exp z\beta. \tag{45}$$

The log-likelihood is then

$$\log L(\lambda, \rho, \beta) = \sum_{i=1}^{d} [\log \lambda + \log \rho + (\lambda - 1) \log \rho \, t_i + z_i \beta]$$

$$+ \sum_{i=1}^{n} - (\rho t_i)^{\lambda} \exp z_i \beta. \tag{46}$$

The parameters λ, ρ and β are estimated by solving the following equation system:

$$\frac{\partial \log L}{\partial \lambda} = \frac{d}{\lambda} + d \log \rho + \sum_{i=1}^{d} \log t_i - \rho^{\lambda} \sum_{i=1}^{n} t_i^{\lambda} \log (\rho t_i) \exp z_i \beta = 0$$

$$\frac{\partial \log L}{\partial \rho} = \frac{\lambda d}{\rho} - \lambda \rho^{\lambda - 1} \sum_{i=1}^{n} t_i^{\lambda} \exp z_i \beta = 0 \tag{47}$$

$$\frac{\partial \log L}{\partial \beta_j} = \sum_{i=1}^{d} z_{ij} - \sum_{i=1}^{n} z_{ij} (\rho t_i)^{\lambda} \exp z_i \beta = 0.$$

In this case, it is only through the Newton–Raphson method, which implies calculating the second derivative of the log-likelihood, that an $\hat{\lambda}, \hat{\rho}$ and $\hat{\beta}$ estimation can be carried out. Therefore, we must calculate the matrix with the following terms:

$$-\frac{\partial^2 \log L}{\partial \lambda^2} = \frac{d}{\lambda^2} + \sum_{i=1}^{n} (\rho t_i)^{\lambda} [\log (\rho t_i)]^2 \exp z_i \beta$$

$$-\frac{\partial^2 \log L}{\partial \rho^2} = \frac{\lambda d}{\rho^2} + \lambda (\lambda - 1) \rho^{\lambda - 2} \sum_{i=1}^{n} t_i^{\lambda} \exp z_i \beta$$

$$-\frac{\partial^2 \log L}{\partial \beta_j^2} = \sum_{i=1}^{n} z_{ij}^2 (\rho t_i)^{\lambda} \exp z_i \beta$$

$$-\frac{\partial^2 \log L}{\partial \lambda \partial \rho} = -\frac{d}{\rho} + \frac{1}{\rho} \sum_{i=1}^{n} [1 + \lambda \log (\rho t_i)] (\rho t_i)^{\lambda} \exp z_i \beta$$

$$-\frac{\partial^2 \log L}{\partial \lambda \partial \beta_j} = \sum_{i=1}^{n} z_{ij} (\rho t_i)^\lambda \log (\rho t_i) \exp z_i \beta$$

$$-\frac{\partial^2 \log L}{\partial \rho \partial \beta_j} = \frac{\lambda}{\rho} \sum_{i=1}^{n} z_{ij} (\rho t_i)^\lambda \exp z_i \beta$$

$$-\frac{\partial^2 \log L}{\partial \beta_j \partial \beta_k} = \sum_{i=1}^{n} z_{ij} z_{ik} (\rho t_i)^\lambda \exp z_i \beta. \tag{48}$$

By considering again the case presented in Section 7.2.1, we are examining the possibility of becoming a home-owner between the ages of 30 and 45 years, for the women of the 'Triple Biographie' survey, according to whether they have no diploma, or at least the Certificat d'études. This estimation provides the following parameters:

$$\hat{\lambda} = 1.4816$$

$$\hat{\rho} = 0.0475$$

$$\hat{\beta}_1 = 0.352.$$

Parameter β_1, being positive, shows that the probability of a woman becoming a home-owner after the age of 30 is stronger when she has a diploma. To determine whether its effect can be considered as significantly different from zero, we must use the matrix of variances and covariances, which has been estimated along with the parameters. In this case, it is written

$$10^{-6} \begin{pmatrix} 1812.60 & 23.07 & 99.90 \\ 23.07 & 2.55 & -69.27 \\ 99.90 & -69.27 & 4019.74 \end{pmatrix}.$$

We can then calculate a χ^2 statistic:

$$\frac{\hat{\beta}_1^2}{V(\hat{\beta}_1)} = \frac{0.124}{4019.74 \times 10^{-6}} = 30.865,$$

which clearly indicates that the effect of the educational level on the probability of becoming a home-owner after the age of 30 is really significant.

Figure 8.2 reports the hazard rates observed and estimated using the Weibull model with regard to the absence or presence of a diploma.

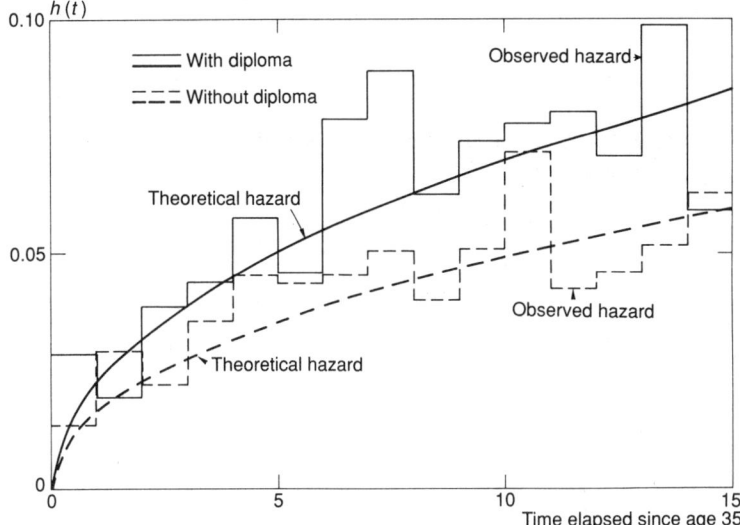

FIG. 8.2. Hazard rates, estimated non-parametrically or using a Weibull model, for the probability of a woman between the ages of 30 and 45 becoming a home-owner according to whether she has no diploma or at least the Certificat d'études

8.3.4. The Gompertz model

Again, we present the estimation of the parameters of a Gompertz model before taking into account various individual characteristics. In this case, the log-likelihood is written

$$\log L(\lambda, \rho) = \sum_{i=1}^{d} (\log \lambda + \log \rho + \rho t_i) + \sum_{i=1}^{n} \lambda [1 - \exp(\rho t_i)] . \quad (49)$$

The first derivatives in relation to the parameters are simply

$$\frac{\partial \log L}{\partial \lambda} = \frac{d}{\lambda} + \sum_{i=1}^{n} (1 - \exp \rho t_i) = \frac{d}{\lambda} + n - \sum_{i=1}^{n} \exp \rho t_i = 0 \quad (50)$$

$$\frac{\partial \log L}{\partial \rho} = \frac{d}{\rho} + \sum_{i=1}^{d} t_i - \lambda \sum_{i=1}^{n} t_i \exp \rho t_i = 0. \quad (51)$$

This equation system is easily solved by substitution. Let us express λ in relation to ρ from the first equation:

$$\lambda = \frac{d}{\sum_{i=1}^{n} \exp \rho t_i - 1}. \tag{52}$$

This value, once transferred to the second, gives an equation where only ρ appears:

$$f(\rho) = \frac{d}{\rho} + \rho \sum_{i=1}^{d} t_i - d \frac{\sum_{i=1}^{d} t_i \exp \rho t_i}{\sum_{i=1}^{d} \exp \rho t_i - 1} = 0. \tag{53}$$

This equation can be solved through successive approximations by using its derivative in relation to ρ:

$$f'(\rho) = -\frac{d}{\rho^2} + \sum_{i=1}^{d} t_i$$

$$- d \frac{\left(\sum_{i=1}^{n} t_i^2 \exp \rho t_i\right)\left(\sum_{i=1}^{n} \exp \rho t_i - 1\right) - \left(\sum_{i=1}^{n} t_i \exp \rho t_i\right)^2}{\left(\sum_{i=1}^{n} \exp \rho t_i - 1\right)^2}. \tag{54}$$

The estimation method is the same as that given in the case of the Weibull model and leads to $\hat{\rho}$ and $\hat{\lambda}$ estimations. Let us consider the second derivative of the opposite of the log-likelihood:

$$-\frac{\partial^2 \log L}{\partial \lambda^2} = \frac{d}{\lambda^2} \tag{55}$$

$$-\frac{\partial^2 \log L}{\partial \rho^2} = \frac{d}{\rho^2} + \lambda \sum_{i=1}^{n} t_i^2 \exp \rho t_i \tag{56}$$

$$-\frac{\partial^2 \log L}{\partial \lambda \, \partial \rho} = \sum_{i=1}^{n} t_i \exp \rho t_i. \tag{57}$$

The estimations of $\hat{\lambda}$ and $\hat{\rho}$ being available, we can estimate the matrix of variances and covariances of these parameters and test various expected values for them. The matrix of variances and covariances can be written

$$\frac{\hat{\lambda}^2 \hat{\rho}^2}{d^2 + d\hat{\lambda}\hat{\rho}^2 \sum_{i=1}^{n} t_i^2 \exp \rho t_i - \hat{\lambda}^2 \hat{\rho}^2 \left(\sum_{i=1}^{n} t_i \exp \rho t_i\right)^2}$$

$$\times \begin{pmatrix} \dfrac{d}{\rho^2} + \lambda \sum_{i=1}^{n} t_i^2 \exp \rho t_i & -\sum_{i=1}^{n} t_i \exp \rho t_i \\ -\sum_{i=1}^{n} t_i \exp \rho t_i & \dfrac{d}{\hat{\lambda}^2} \end{pmatrix} \qquad (58)$$

Let us now introduce the effect of the various characteristics. In the proportional hazard model, there was

$$h(t; z) = \lambda \rho \exp (\rho t + z\beta). \qquad (59)$$

The log-likelihood is then written

$$\log L (\lambda, \rho, \beta) = \sum_{i=1}^{d} (\log \lambda + \log \rho + \rho t_i + z_i \beta)$$

$$+ \sum_{i=1}^{n} \lambda (1 - \exp \rho t_i) \exp z_i \beta. \qquad (60)$$

The derivative in relation to the parameters are written, in order to simplify $\log L (\lambda, \rho, \beta) = l$,

$$\frac{\partial l}{\partial \lambda} = \frac{d}{\lambda} + \sum_{i=1}^{n} (1 - \exp \rho t_i) \exp z_i \beta = 0$$

$$\frac{\partial l}{\partial \rho} = \frac{d}{\rho} + \sum_{i=1}^{d} t_i - \lambda \sum_{i=1}^{n} t_i (\exp \rho t_i) \exp z_i \beta = 0 \qquad (61)$$

$$\frac{\partial l}{\partial \beta_j} = \sum_{i=1}^{d} z_{ij} + \lambda \sum_{i=1}^{n} z_{ij} (1 - \exp \rho t_i) \exp z_i \beta = 0.$$

This system is solved using the Newton–Raphson method, which implies the calculation of minus the second derivative:

$$-\frac{\partial^2 l}{\partial \lambda^2} = \frac{d}{\lambda^2}$$

$$-\frac{\partial^2 l}{\partial \rho^2} = \frac{d}{\rho^2} + \lambda \sum_{i=1}^{n} t_i^2 (\exp \rho t_i) \exp z_i \beta$$

$$-\frac{\partial^2 l}{\partial \beta_j^2} = -\lambda \sum_{i=1}^{n} z_{ij}^2 (1 - \exp \rho t_i) \exp z_i \beta$$

$$-\frac{\partial^2 l}{\partial \lambda \partial \rho} = \sum_{i=1}^{n} t_i (\exp \rho t_i) \exp z_i \beta$$

$$-\frac{\partial^2 l}{\partial \lambda \partial \beta_j} = -\sum_{i=1}^{n} z_{ij} (1 - \exp \rho t_i) \exp z_i \beta$$

$$-\frac{\partial^2 l}{\partial \rho \partial \beta_j} = \lambda \sum_{i=1}^{n} t_i z_{ij} (\exp \rho t_i) \exp z_i \beta$$

$$-\frac{\partial^2 l}{\partial \beta_j \partial \beta_k} = -\lambda \sum_{i=1}^{n} z_{ij} z_{ik} (1 - \exp \rho t_i) \exp z_i \beta. \tag{62}$$

The RATE program, written by N. Tuma and D. Pasta, makes it possible to estimate all these parameters. In this case, the model is written slightly differently:

$$h(t; z) = \exp(\rho t + z\beta). \tag{63}$$

In fact, the first vector z_1 always takes value 1, which implies that

$$\exp(\beta_1) = \lambda \rho. \tag{64}$$

All other parameters are identical to the model presented here.

As an example, we give the results obtained when studying the duration of stay in a dwelling for men born between 1931 and 1935. These data are again taken from the 'Triple Biographie' survey.[3] Here, we consider the age of the individual at the beginning of the stay, which is taken account of in the form of a series of binary variables: under 20, between 20 and 24, between 25 and 29, between 30 and 34, between 35 and 39, and between 40 and 44. To make the estimation, one must not take account of the group aged 45 years and over, which is already defined when all previous binary variables are nil.

The number of durations of stay observed is 2523, of which 493 are still ongoing at the time of the survey.

For a comparison, we choose the exponential model in which no variable is taken into account. This model leads to a constant hazard rate estimated at 0.1237 with -6273.15 as maximum value of the log-likelihood.

---

[3] For further details, see Courgeau (1985a, b).

TABLE 8.1. Estimation of the parameters β, the standard deviation, the χ^2 test with one degree of freedom of the nullity of their value, and $\exp(\beta_i)$

Characteristic considered	β_i	$\sigma(\hat{\beta}_i)$	$\left[\dfrac{\hat{\beta}_i}{\sigma(\hat{\beta}_i)}\right]^2$	$\exp \beta_i$
Constant	−2.727	0.2880	89.187	0.0654
Under 20 years	1.131	0.2920	15.008	3.100
20–24 years	1.410	0.2912	23.460	4.097
25–29 years	0.892	0.2940	9.203	2.440
30–34 years	0.626	0.2978	4.417	1.870
35–39 years	0.135	0.3064	0.194	1.145
40–44 years	−0.177	0.3292	0.127	0.889

Taking into account the various age groups in a Gompertz model provides a new maximum of the log-likelihood equal to −5991.67. Using the logarithm of the likelihood ratio of the second model compared with the first one, we obtain $-2 \log R = 562.96$, which has a χ^2 distribution with seven degrees of freedom, if the second model brings nothing more than the first. The probability of reaching this value being very low, the second model is far more satisfactory than the first.

An estimation of the various β and ρ parameters is given in Table 8.1, together with their standard deviation and the test of the nullity of their value, which is χ^2 with one degree of freedom if the z_i variable has no effect on the probability of migrating, and lastly the $\exp \beta_i$, values. These values are easy to interpret in the case of binary variables: they indicate the increase or decrease in the hazard rate when the individual has this characteristic.

Parameter $\hat{\rho}$ is equal to −0.0629 with a standard deviation of 4.485×10^{-3}, resulting in a test of χ^2 with one degree of freedom equal to 196.784. This shows that the duration of stay has a strong effect on the probability of migrating. Thus, after a 10-year stay, the probability is almost halved:

$$\exp(-0.629) = 0.525.$$

At the 5 per cent threshold, the age effect is really significant before 35 years. Beyond that age, the probability of migration can be considered as constant and equal to 0.0654, which is half the rate calculated for all the migrations. This age effect is most important between

TABLE 8.2. Matrix of the estimated variances and covariances of various parameters in the Gompertz model

	Constant β_1	Under 20	20–24	25–29	30–34	35–39	40–44	Constant ρ
Constant β_1	8.3387×10^{-2}							
Under 20 years	-8.3240×10^{-2}	8.5282×10^{-2}						
20–24 years	-8.3242×10^{-2}	8.3491×10^{-2}	8.4786×10^{-2}					
25–29 years	-8.3205×10^{-2}	8.3556×10^{-2}	8.3550×10^{-2}	-8.6439×10^{-2}				
30–34 years	-8.3219×10^{-2}	8.3533×10^{-2}	8.3527×10^{-2}	8.3606×10^{-2}	8.8706×10^{-2}			
35–39 years	-8.3240×10^{-2}	8.3495×10^{-2}	8.3490×10^{-2}	8.3555×10^{-2}	8.3531×10^{-2}	8.3911×10^{-2}		
40–44 years	-8.3284×10^{-2}	8.3420×10^{-2}	8.3417×10^{-2}	8.3452×10^{-2}	8.3439×10^{-2}	8.3419×10^{-2}	1.0838×10^{-1}	
Constant ρ	-3.2884×10^{-5}	-5.7207×10^{-5}	-5.5561×10^{-5}	-7.8316×10^{-5}	-7.0116×10^{-5}	-5.6831×10^{-5}	-3.0415×10^{-5}	-2.0112×10^{-5}

Estimation of Parametric Models

the ages of 20 and 24, when the probability of migration is more than four times that of individuals aged 35 years and over.

The variances and covariances matrix of the β and ρ parameters are given in Table 8.2. These variances and covariances can be used to carry out various tests. To test whether the probability of migration is different at ages 35–39 and 40–44, for instance, we can build the statistic $\beta_1' C_1 \beta_1$ where $\beta_1' = (0.135 - 0.117)$, β_1 is the corresponding column vector, and C_1 is the following matrix of variances and covariances:

$$C_1 = \begin{pmatrix} 9.3911 \times 10^{-2} & 8.3419 \times 10^{-2} \\ 8.3419 \times 10^{-2} & 1.0838 \times 10^{-1} \end{pmatrix}.$$

If the two parameters cannot be considered as different, this statistic is a χ^2 with two degrees of freedom. In that case the result is 0.056, which shows that the two age groups can be combined in a single one.

8.3.5. Log-logistic model with accelerated failure times

This last example corresponds to the case where the hazard rate can take a maximum value before decreasing.

We start directly from the model which takes into account various characteristics of the individuals. In Chapter 7 (formula (111)) we showed that the hazard rate is written

$$h(t; \lambda, z) = \frac{\lambda}{t\,[\,1 + t^{-\lambda} \exp(-z\beta)\,]}, \qquad (65)$$

where parameter ρ has been introduced using a first variable z_1 equal to the unit for all respondents, which means that

$$\exp(\beta_1) = \rho^\lambda. \qquad (66)$$

In that case, the survivor function is written

$$S(t; \lambda, z) = [\,1 + t^\lambda \exp(z\beta)\,]^{-1}. \qquad (67)$$

The log-likelihood then becomes

$$\log L(\lambda, z) = \sum_{i=1}^{d} \{\log \lambda - \log t_i - \log[1 + t_i^{-\lambda} \exp(-z_i \beta)]\}$$

$$- \sum_{i=1}^{n} \log[1 + t_i^\lambda \exp(z_i \beta)] \qquad (68)$$

170   Extending the Scope of Regression Models

The derivative in relation to the parameters are written

$$\frac{\partial l}{\partial \lambda} = \frac{d}{\lambda} + \sum_{i=1}^{d} \frac{\log t_i}{1 + t_i^{\lambda} \exp(z_i \beta)} - \sum_{i=1}^{n} \frac{\log t_i}{1 + t_i^{-\lambda} \exp(-z_i \beta)} = 0$$

$$\frac{\partial l}{\partial \beta_j} = \sum_{i=1}^{d} \frac{z_{ij}}{1 + t_i^{\lambda} \exp(z_i \beta)} - \sum_{i=1}^{n} \frac{z_{ij}}{1 + t_i^{-\lambda} \exp(-z_i \beta)} = 0.$$

(69)

Again, the Newton–Raphson method can be used to solve this system.

Calculating the opposite of the second derivative in relation to the parameters gives

$$-\frac{\partial^2 l}{\partial \lambda^2} = \frac{d}{\lambda^2} + \sum_{i=1}^{n} \frac{2^{\delta_i} (\log t_i)^2}{[1 + t_i^{-\lambda} \exp(-z_i \beta)] [1 + t_i^{\lambda} \exp(z_i \beta)]}$$

$$-\frac{\partial^2 l}{\partial \beta_j^2} = \sum_{i=1}^{n} \frac{2^{\delta_i} z_{ij}^2}{[1 + t_i^{-\lambda} \exp(-z_i \beta)] [1 + t_i^{\lambda} (\exp z_i \beta)]}$$

$$-\frac{\partial^2 l}{\partial \lambda \partial \beta_j} = \sum_{i=1}^{n} \frac{2^{\delta_i} z_{ij} \log t_i}{[1 + t_i^{-\lambda} \exp(-z_i \beta)] [1 + t_i^{\lambda} \exp(z_i \beta)]}$$

$$-\frac{\partial^2 l}{\partial \beta_j \partial \beta_k} = \sum_{i=1}^{n} \frac{2^{\delta_i} z_{ij} z_{ik}}{[1 + t_i^{-\lambda} \exp(-z_i \beta)] [1 + t_i^{\lambda} \exp(z_i \beta)]},$$

(70)

where $\delta_i = 1$ when the event occurs before the censoring time of the i th individual and $\delta_i = 0$ in the opposite case. The inverse of this information matrix enables us to estimate $\hat{\lambda}$ and the values of $\hat{\beta}_0$. For these values of the parameters, it also provides an estimation of the matrix of their variances and covariances.

Let us apply this method to the fact of becoming a home-owner, in relation to the women's age, starting from 15. At first, we do not take any characteristic into account. Using a value $\lambda^* = 2$ and $\beta_1^* = -6$, we obtain the maximum of the log-likelihood, after five iterations, with a value of inferior derivative of 10^{-3}. The values obtained are

$$\hat{\lambda} = 1.922$$

$$\hat{\beta}_1 = 6.556$$

that is, $\rho = 0.0341$. The log-likelihood, which was equal to -7463.34 for the initial values, is now equal to -7146.79. The matrix of variances and covariances is estimated as

Estimation of Parametric Models

$$10^{-4}\begin{pmatrix} 17.47 & -57.24 \\ -57.24 & 200.99 \end{pmatrix}.$$

Figure 8.3 reports the hazard rates calculated using the non-parametric method, along with those obtained using the log-logistic model. It appears that the model corresponds to the observations between the ages of 15 and 35 years, that it underestimates them between 35 and 45, and that it overestimates them later.

Figure 8.4 also reports the survivor functions in the state of non-ownership, calculated non-parametrically and using the log-logistic model. For the non-parametric model, it appears that working with the survivor function masks the random differences revealed with the hazard rates. However, the non-parametric survivor function is beyond that of the log-logistic model for women over 35, but it exceeds it again after 55.

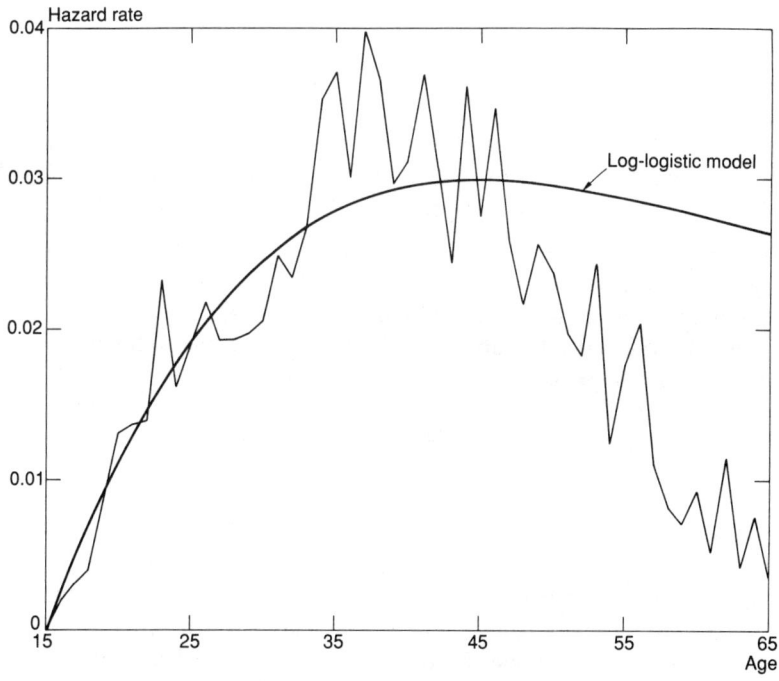

FIG. 8.3. Hazard rates corresponding to the fact of becoming a home-owner for women, from 15 years of age onwards, estimated non-parametrically or using a log-logistic model

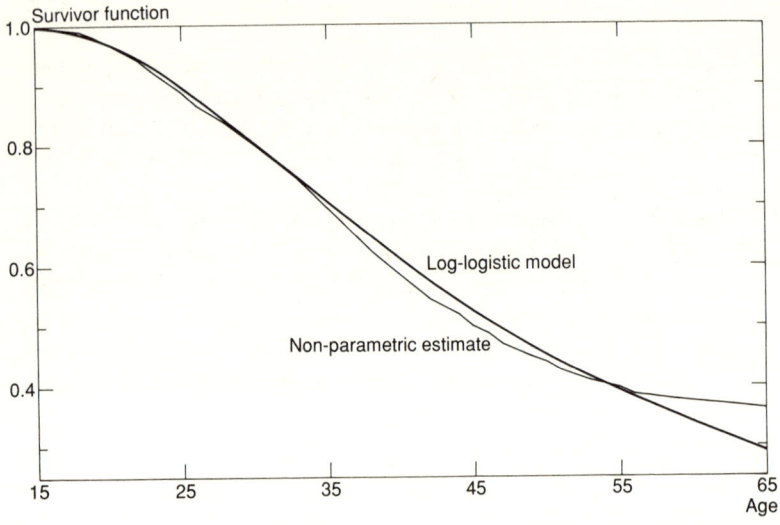

FIG. 8.4. Survivor function for becoming a home-owner for women after the age of 15, estimated non-parametrically or using a log-logistic model

Let us now introduce the educational level of the respondents in the form of a binary variable z_2, equal to 0 when the individual has no diploma, equal to 1 when he has at least the Certificat d'études. Applying the Newton–Raphson technique leads to the following estimation of the parameters:

$$\hat{\lambda} = 1.928$$

$$\hat{\beta}_1 = -6.732$$

$$\hat{\beta}_2 = 0.256;$$

that is, $\rho = 0.0019$. Parameter $\hat{\beta}_2$, being positive, shows that there is again a stronger probability of becoming a home-owner after the age of 15 when a woman has a diploma. To determine whether it has a significant effect different from zero, we must use the matrix of variances and covariances, which has been estimated along with these parameters. In that case, it is written

Estimation of Parametric Models

$$10^{-4}\begin{pmatrix} 17.58 & -58.60 & 1.57 \\ -58.60 & 230.75 & -40.69 \\ 1.57 & -40.69 & 57.22 \end{pmatrix}.$$

We can therefore compute a statistic of χ^2:

$$\frac{\hat{\beta}_2^2}{V(\hat{\beta}_2)} = \frac{0.0655}{57.22 \times 10^{-4}} = 11.45,$$

showing that the effect of the educational level on the probability of becoming a home-owner is really significant at all ages, as it is for women over 30. This confirms the differences that have been revealed in Fig. 7.6 above.

8.4. Comparison of Parametric Models

First, it is possible to introduce a great number of variables in all these models. The data used are often taken from biographical surveys, so that many elements of the respondents' lives are available and can be linked to the event under study. Therefore, it is important to choose, from among the models that take these different variables into account, the one that best explains the phenomenon.

As an example, let us consider again the results obtained when studying the durations of stay in a dwelling for men born between 1931 and 1935. Earlier, we presented the results obtained by modifying the age of the individual at the beginning of the stay and the duration of the stay. The 'Triple Biographie' survey provides many other elements concerning the respondents' biography and characteristics. In this case, it is important to take them into account and to decide whether this improves the quality of the model. Thus, the variables for the stage of the family life-cycle that the respondent has reached can influence the probability of his migration; for example, does an individual who is single at the outset migrate in a different way from one who is married? The status of occupation of the dwelling can also influence this mobility. Similarly, the stages that the individual goes through in his professional life can make him more stable or, on the other hand, can induce him to migrate. Political events (wars, military service, etc.) might also entail particular mobilities. Lastly, elements of the family origin (mobility during childhood, number of brothers and sisters, etc.) can influence mobility.[4]

---

[4] For more details, see Courgeau (1985*a*, *b*).

TABLE 8.3. The effect of adding new variables on the accuracy of the model to study the changes of dwelling of men born between 1930 and 1935

Type of model	New added variables	No. of added variables	Maximum of the log-likelihood	Diff. of χ^2 in relation to previous model
Exponential	Constant	1	−6273.15	—
Exponential	Age groups	6	−6113.38	319.53
Gompertz	Duration of stay	1	−5991.67	243.43
Gompertz	Family charact.	5	−5963.17	57.00
Gompertz	Residential status	3	−5755.45	415.44
Gompertz	Professional charact.	10	−5685.59	139.72
Gompertz	Political events	3	−5642.93	85.31
Gompertz	Family origins	3	−5637.67	10.53

We have chosen to introduce these elements cumulatively, starting from an exponential model and working towards a Gompertz model. Thus, we can see whether adding new variables improves the quality of the model. This quality is measured by the log-likelihood ratio between successive models.

The result of these different stages is reported in Table 8.3. All χ^2 are significant, which implies that adding new variables brings new elements to explain the migration. The migratory behaviour pattern is perhaps best explained by some variables that have been introduced along the way and are correlated with the age group. In this case, the impact of age would be reduced in models that take more particular account of the matrimonial status, the status of occupation, the dwelling, etc. In fact, we see that a married individual's mobility is reduced by 80 per cent compared with that of a single person; that a home-owner's mobility is reduced by 20 per cent compared with that of a tenant; etc. This is what we observe when all variables have been introduced into the model.

Table 8.4 reports the effect of belonging to different age groups, once all other characteristics have been taken account of. It clearly indicates that this does not influence the probability of migration. On the other hand, the effect of precise characteristics, such as the matrimonial status, the fact of being a home-owner, or of having had very mobile parents during childhood, etc., far better explain the changes in the migratory behaviour patterns of the individual.

Estimation of Parametric Models

TABLE 8.4. The effect of belonging to a certain age group at the beginning of stay when all variables are introduced simultaneously into the model

Age groups	$\hat{\beta}_i$	$\sigma(\hat{\beta}_i)$	$\left[\dfrac{\hat{\beta}_i}{\sigma(\hat{\beta}_i)}\right]^2$	$\exp \beta_i$
Under 20 years	0.234	0.298	0.619	1.264
20–24 years	0.201	0.295	0.466	1.223
25–29 years	0.014	0.294	0.002	1.014
30–34 years	−0.160	0.297	0.293	0.852
35–39 years	−0.399	0.303	1.741	0.671

To compare various models and to choose the most adequate, we can also use a model general enough to cover the greatest number of models presented here, as well as particular cases. The previous chapter has already presented the generalized Fisher–Snedecor distribution which provides this opportunity.

When all events have been observed, the log-likelihood of this distribution is written

$$\log L(\rho, \sigma, k_1, k_2) = \sum_{i=1}^{n} \Big[k_1 \log k_1 + k_2 \log k_2 + \log \Gamma(k_1 + k_2)$$

$$- \log \Gamma(k_1) - \log \Gamma(k_2) + \frac{k_1}{\sigma} \log \rho + \left(\frac{k_1}{\sigma} - 1\right) \log t_i$$

$$- \log \sigma - (k_1 + k_2) \log \big(k_2 + k_1 (\rho t_i)^{1/\sigma}\big) \Big], \quad (71)$$

resulting in the derivatives in relation to various parameters:

$$\frac{\partial l}{\partial \rho} = \sum_{i=1}^{n} \frac{k_1}{\sigma \rho} \left(1 - \frac{k_1 + k_2}{\sigma \rho \, [k_2 (\rho t_i)^{-1/\sigma} + k_1]} \right) = 0$$

$$\frac{\partial l}{\partial \sigma} = \sum_{i=1}^{n} \left(-\frac{k_1}{\sigma^2} \log (\rho t_i) - \frac{1}{\sigma} + (k_1 + k_2) \frac{k_1 \log \rho t_i}{\sigma^2 [k_2 (\rho t_i)^{-1/\sigma} + k_1]} \right) = 0$$

$$\frac{\partial l}{\partial k_1} = \sum_{i=1}^{n} \Big(\log k_1 + 1 + \psi(k_1 + k_2) - \psi(k_1) + \frac{\log \rho t_i}{\sigma}$$

$$- \log [k_2 + k_1 (\rho t_i)^{-1/\sigma}] - \frac{k_1 + k_2}{k_2 (\rho t_i)^{-1/\sigma} + K_1} \Big) = 0$$

$$\frac{\partial l}{\partial k_2} = \sum_{i=1}^{n} \left(\log k_2 + 1 + \psi(k_1 + k_2) - \psi(k_2) + \log(k_2 + k_1 (\rho t_i)^{1/\sigma}) \right.$$

$$\left. - \frac{k_1 + k_2}{k_2 + k_1 (\rho t_i)^{1/\sigma}} \right) = 0, \qquad (72)$$

where the function $\psi(k) = \partial \log \Gamma(k) / \partial k$ is the derivative of function Γ. Similarly, it is possible, by differentiating these expressions anew, to obtain the information matrix[5] and use the Newton–Raphson technique to estimate the four parameters of the model. However, to estimate the parametric values of the model when k or k_2 tend towards the infinite and to test these values, it is useful to replace k_1 and k_2 with two new parameters q_1 and q_2, equal to

$$q_1 = \frac{k_2 - k_1}{\sqrt{k_1 k_2 (k_1 + k_2)}}$$

$$q_2 = \frac{2}{k_1 + k_2}, \qquad (73)$$

which gives for k_1 and k_2, in relation to these new parameters,

$$k_1 = \frac{2}{q_1^2 + 2q_2 + q_1 \sqrt{q_1^2 + 2q_2}} \qquad (74)$$

$$k_2 = \frac{2}{q_1^2 + 2q_2 - q_1 \sqrt{q_1^2 + 2q_2}} \qquad (75)$$

With these new parameters, the function of the log-likelihood is finite when k_1 or k_2 tend towards the infinite, and the system of equations (72) has a solution. Thus, for instance, if $k_1 \to \infty$, we can verify that $q_1 \to 1/\sqrt{k_2}$ and $q_2 \to 0$.

The log-normal model corresponds to point (0,0), the Weibull model corresponds to point (1,0), and the log-logistic model to point (0,1). This family includes other kinds of models as well. It is therefore possible to test the validity of the different types of models in the general case.[5]

If the possibility of a censoring before the individual experiences the event is now introduced, the calculations quickly become complex, for there is no simple expression for hazard rates or the survivor

---

[5] For more details on this estimation and the possible tests, see Prentice (1975).

function. However, if losses are few, it is always possible to approach the likelihood through the relation (71), which provides a rough estimate of the parameters of the model.

8.5. Conclusion

This chapter has presented some precise methods for estimating the parameters of various models discussed in the previous chapter, and for measuring the effect of characteristics that can influence the probability of experiencing the event under study. In each case, we have indicated the equations necessary for the estimation along with the best way to solve them.

At the same time, these methods provide an estimation of the matrix of variances and covariances of the parameters estimated. Using them, all desirable tests can be carried out, showing in particular which characteristics have a significant impact on the duration of stay.

Throughout this chapter, we have worked on distributions that are continuous functions of time. However, in some surveys the data are grouped together for given periods (on a quarterly or annual basis, for example). Using the models presented here, it is easy to generate the corresponding discrete-time models by introducing a classification according to time. For example, in the case of annual data, we can introduce an underlying continuous time T', and the time measured T represents the part of this underlying time which is whole. In that case, we can write

$$f(t) = P(T = t) = P(t \leq T' < t+1) = S(t) - S(t+1) \qquad (76)$$

$$\lambda(t) = P(T' < t+1 \mid T' \geq t) = \frac{S(t) - S(t+1)}{S(t)}. \qquad (77)$$

With these notations, the likelihood of observations can be expressed in a similar way to those in the continuous case (see formula (4)):

$$\log L(\beta) = \sum_{i=1}^{n} \left[\delta_i \log \left(1 - \frac{S(t_i + 1; z_i, \beta)}{S(t_i, z_i, \beta)} \right) + \log S(t_i, z_i, \beta) \right]. \qquad (78)$$

The methods of estimation presented are still valid in that case. Note however that, when the grouping interval is short in relation to the whole period of observation, using a discrete-time model gives results almost identical to those of a continuous-time model.

All characteristics that have been introduced are assumed to be defined at the beginning of the stay. Actually, they can evolve before the occurrence of the event under study and can change its probability of occurrence. Thus, when studying the probability of becoming a home-owner, this probability can be modified by the fact of becoming married. It is useful, therefore, to introduce time-dependent characteristics. Thus, the matrimonial status variable can take value 0 until the marriage, when it becomes 1. The methods presented here can easily be generalized in that case, although the calculations are very difficult to carry out. In fact, the log-likelihood (5), when all z variables depend on time, can be written

$$\log L(\beta) = \sum_{i=1}^{n} \left[\delta_i \log h(t_i; z_i(t_i), \beta) + \int_0^{t_i} h(t; z_i(t), \beta) dt \right]. \quad (79)$$

Searching for the values of β which maximize that function is done in a form identical to that presented in this chapter. It is also possible to introduce an unobserved heterogeneity, if we have some knowledge of its distribution among the respondents. Chapter 7 included a presentation of a certain number of distributions obtained by introducing an unobserved heterogeneity, in the case of an exponential overall model. Again, these models can be estimated using the methods presented in this chapter. Note however that in social sciences, if the effect of many observed characteristics on the respondents has been highlighted, there are as yet few elements on the differences in behaviours that are independent of these characteristics. Therefore, in the absence of reliable information, modelizing this unobserved heterogeneity constitutes a great risk. Heckman and Singer (1982) have shown that, according to the chosen distribution, the effect of some characteristics can vary to a large extent. Trussel and Richards (1985) have shown that the conclusions of the analysis depend not only on the distributions chosen, but also on the type of dependence on time of the variables introduced. This unobserved heterogeneity can be largely reduced when a maximum number of characteristics of the respondents have been taken into account. This prevents an inaccurate modelization of the unobserved heterogeneity.

Lastly, the use of parametric models is subjected to very strong hypotheses on the distribution of hazard rates, and the number of respondents available is often too small to verify them with precision. To avoid this, it is possible to consider more general models, which take parametrically into account only the effect of characteristics, with a non-parametric estimation of the underlying instantaneous failure rates. In particular, proportional hazard models offer this possibility. This semi-parametric analysis is developed in the next chapter.

9

Methods of Semi-parametric Analysis

In event history analysis, semi-parametric methods are preferred to parametric methods by many researchers, not least because they make it possible to relate hazard rates to individual characteristics without imposing a formalization of the duration effect. We shall therefore discuss in detail the contribution of these methods.

9.1. From Parametric Regressions to Semi-parametric Proportional Hazard Models

To analyse the relations between explanatory variables and the occurrence of an event, we have presented many parametric formalizations of the distribution of the event—particularly the exponential, Weibull, or Gompertz distributions, which differ in the way in which time is taken into account, but also other log-normal or gamma distributions, which are more difficult to estimate at a practical level.

As we saw, these models take account of the variables with a multiplicative effect on hazard rates (property of the *proportional hazard models*). They also define another class of log-linear models called *accelerated failure time models* because of the multiplicative effect the covariates have on T.

9.1.1. Definition

The semi-parametric models, introduced by Cox (1972), modelize the hazard rates in the following manner:

$$h(t; z) = h_0(t) \exp(\beta z). \qquad (1)$$

But this time $h_0(t)$ is an unknown arbitrary function of t called a *baseline hazard*.

These models therefore constitute a generalization of the previous ones; for example, if $h_0(t) = \exp(a)$, then the model is exponential.

When interpreting, $h_0(t)$ is most often the instantaneous hazard rate of the standard individual ($z = 0$). However, as h_0 is not defined, the model remains partially unspecified, and is therefore called semi-parametric. On the other hand, since the incorporated variables are fixed, its being called a proportional hazard model originates in the relation that exists between the conditional densities of two individuals. Since the ratio of the densities of two individuals is constant whatever t is, their hazard rates are thus really proportional (see Section 7.2.1).

9.1.2. Building the likelihood

Let us consider a sample of n individuals. The risk at time t is measured by the hazard $h(t; z)$. We want to estimate the unknown β parameters which measure the influence of z on $h(\cdot)$. For each of i individuals, there are t_i, c_i, z_i, where t_i is the failure time of the event if $\delta_i = 1$, where c_i is the censoring time (loss of the individual to the follow-up) if $\delta_i = 0$. The likelihood is formed as follows:

$$L = \prod_{i=1}^{n} [f(t_i; z_i)^{\delta_i} S(t_i; z_i)^{1-\delta_i}], \tag{2}$$

Once developed, this is written

$$L = \prod_{i=1}^{n} h(t_i; z_i)^{\delta_i} \exp\left[-\int_0^\infty \sum_{l \in R_{t_i}} h(u; z_l) du\right], \tag{3}$$

where R_t is the population at risk in $t - 0$, an expression similar to (4) in Section 8.1. As part of the expression of the hazard rates is unspecified, i.e. the baseline hazard $h_0(t)$, the various parameters cannot be estimated by a direct maximization of the above-mentioned likelihood. In fact, in the absence of constraints on $h_0(t)$, it is not possible to find a maximum to the expression of likelihood. Therefore, we shall use specific estimation methods.

It is in fact necessary to estimate the value of the parameters β that modify parametrically the hazard rate $h(t, z)$. This estimate is measured by maximizing a partial form of the likelihood. Then, knowing the value of $\hat{\beta}$, we estimate the baseline hazard $h_0(t)$ non-parametrically.

9.2. Methods of Estimation

9.2.1. Estimating the parameters

Let us first show the close relationship between the marginal likelihood and the partial likelihood, which we maximize to obtain the estimators of the parameters.

The expression of marginal likelihood derives from its elaboration in relation to the marginal distribution of ranks and rank-testing techniques (Kalbfleisch and Prentice, 1980). Let $t_1, \ldots, t_n$ be the failure times for the n individuals of the sample with the corresponding covariates $z_1, \ldots, z_n$. By forming $t_{(1)} < t_{(2)} < \ldots < t_{(n)}$, series of the ordered times, we obtain two new statistics:

$$\theta(t) = [t_{(1)}, \ldots, t_{(n)}], \text{ the order statistic}$$

and

$$R(t) = [(1), \ldots, (n)], \text{ the rank statistic.}$$

The first of these contains the ordered times and the second refers to the corresponding label attached to the ordered times. For instance, if $n = 4$ and we observe $t_1 = 12$, $t_2 = 5$, $t_3 = 9$, $t_4 = 17$, then

$$\theta(t) = [5, 9, 12, 17] \quad \text{and} \quad R(t) = [2, 3, 1, 4].$$

Let G be the group of transformations of $\mathbb{R}^+$ in $\mathbb{R}^+$ strictly increasing and differentiable. The problem of estimating the parameters β in the expression of the hazard rates $h(t; z) = h_0(t) \exp(\beta z)$ remains untransformed with regard to the group G of the transformations applied to t. Moreover, the action of these transformations over $\theta(t)$ leaves $R(t)$ unchanged. In the above example, if the transformation is defined by $u = 3 \times t$, then

$$\theta(u) = [15, 27, 36, 51] \quad \text{and} \quad R(u) = [2, 3, 1, 4] = R(t).$$

We may therefore consider that the problem of estimating the parameters β is the same whatever the transformation imposed to $\theta(t)$ and that only $R(t)$ contains the information on the parameters β when $h_0(t)$ is completely unknown. Under these conditions, the order statistic $R(t)$ is called marginally sufficient for the estimation of the β when $h_0(t)$ is unspecified. Thus, the order of the occurrences is more important than the exact failure times.

Methods of Semi-parametric Analysis

We then form a marginal likelihood which is proportional to the probability of observing the chronology as it has been collected. This amounts to calculating the partial likelihood according to the definition of Cox. Conditionally on the population at risk and on the fact that an event occurs at t_i, the probability that an individual i experiences the event is equal to

$$\frac{h_0(t_i) \exp(z_i(t_i)\beta)}{\sum_{l \in R_i} h_0(t_i) \exp(z_l(t_i)\beta)}, \qquad (4)$$

where R_i represents all the labels of the individuals at risk in $t_i - 0$. This expression comes down to

$$\frac{\exp(z_i(t_i)\beta)}{\sum_{l \in R_i} \exp(z_l(t_i)\beta)}. \qquad (5)$$

We thus leave out the intervals where no event occurs and for which no information on the z is available. The partial likelihood is then formed as the product on all failure times:

$$PL(B) = \prod_{i=1}^{n} \frac{\exp(z_i(t_i)\beta)}{\sum_{l \in R_i} \exp(z_l(t_i)\beta)}, \qquad (6)$$

which can be written more simply as

$$PL(\beta) = \frac{\exp \sum_{i=1}^{n} z_i(t_i)\beta}{\prod_{i=1}^{n} \left[\sum_{l \in R_i} \exp(z_l(t_i)\beta) \right]}, \qquad (7)$$

In practice, there are several occurrences at the same time within the sample. This possibility must therefore be taken into account. In this case, we observe n individuals and the ordered failure times are $t_1 < \ldots < t_k$. Let us denote by d_i the number of individuals for whom the failure time is t_i. Knowing that the ranks thus defined are not affected by the value of the d_i, we then have

$$PL(\beta) = \prod_{i=1}^{n} \frac{\exp(s_i\beta)}{\sum_{l \in R_i} \left[\exp(z_l\beta) \right]^{d_i}}, \qquad (8)$$

where $s_i = \Sigma z_j(t_i)$ is the sum of the explanatory variables of the d_i individuals who experience the event (or the censoring) in t_i. The estimator β of the maximum likelihood is then obtained as a solution to the current equations:

$$\frac{d \log PL(\beta)}{d\beta_i} = 0; \tag{9}$$

that is,

$$s_{ji} - d_i A_{ji}(\beta) = 0, \tag{10}$$

where s_{ji} is the jth component of the vector s_i and

$$A_{ji}(\beta) = \frac{\sum_{l \in R_i} z_{jl} \exp(z_l \beta)}{\sum_{l \in R_i} \exp(z_l \beta)}. \tag{11}$$

Let us now show the relations between this partial likelihood and the total likelihood. The method of estimation of the partial likelihood as proposed by Cox represents a fundamental contribution. In fact, its implementation is very similar to that of ordinary estimations of the maximum likelihood, in a context where the probability densities associated with the distributions studied are complex. Without discussing the formal details of Cox's demonstration, it is possible to present its principle in relation to the maximum likelihood technique.

The partial likelihood method is very close to that of the maximum likelihood in that it also involves two stages:

1. building a likelihood concerning the unknown parameters based on the observation;
2. finding the values that maximize the function built.

There is, however, a difference. While the total likelihood is the product of the contributions of each individual in the sample, the partial likelihood is the product of the contributions of each event observed.

To build the partial likelihood, we keep only one of the factors of the total likelihood. In fact, in the context of the proportional hazard modelization, the total likelihood consists of two factors: one which contains the information on the parameter β, and another which contains the information concerning β and $h_0(t)$. The partial likelihood, therefore, retains only the first factor, treating it as a full likelihood.

Methods of Semi-parametric Analysis

This first factor depends only on the order of occurrence of events (which makes it possible to show that the partial likelihood is in fact a marginal likelihood) rather than on their exact failure time.

The obtained estimators are asymptotically unbiased and distributed as usual. However, they are not consistent since the information on the exact failure times is not used in the estimation.

9.2.2. Estimating the survivor function

Once the estimation of the parameters associated with the explanatory variables is presented in detail, it is more difficult to estimate the non-parametric element in instantaneous failure rates. These are expressed in the form

$$h(t; z) = h_0(t) \exp(z\beta).$$

The survivor function in a semi-parametric modelization is expressed in the form

$$S(t; z) = \exp - \int_0^t h_0(u) \exp(z\beta) du \qquad (12)$$

or else

$$S(t; z) = \left[\exp - \int_0^t h_0(u) du \right]^{\exp(z\beta)} \qquad (13)$$

and, lastly,

$$S(t; z) = S_0(t)^{\exp(z\beta)}. \qquad (14)$$

Using the partial likelihood method, it is possible to estimate the parameters β without making any hypothesis on the $h_0(\cdot)$.

Let us now estimate the $h_0(\cdot)$ using the estimated parameters β but by applying a method of estimation that is almost non-parametric. Let $t_1 < \ldots < t_k$ be the occurrences observed, and let us assume that during the interval $[t_i, t_{i+1}[$ the losses occur in t_l which belongs to this interval. The likelihood is therefore constituted by the contributions of those who experience the event at the beginning of the interval; that is,

$$S_0(t_i)^{\exp(z\beta)} - S_0(t_i + 0)^{\exp(z\beta)}, \qquad (15)$$

where T_i is the full set of their labels.

The contribution of those who are censored in t_l is

$$S_0(t_l + 0)^{\exp(z\beta)}, \qquad (16)$$

where M_i is the full set of their labels.

It follows that the likelihood is equal to

$$L = \prod_{i=1}^{k} \left[\prod_{l \in T_i} (S_0(t_i)^{\exp(z_l \beta)} - S_0(t_i + 0)^{\exp(z_l \beta)}) \prod_{l \in M_i} S_0(t_l + 0)^{\exp(z_l \beta)} \right]. \qquad (17)$$

As in the case of the Kaplan–Meier estimator, we assume that the events occur at the beginning of the interval:

$$S_0(t) = S_0(t_i + 0) \quad \text{when} \quad t_1 < t \leq t_{i+1}$$

in a discrete-time model where the hazard rates are given as

$$h_0(t_i) = 1 - \alpha_i.$$

Using this notation, we can write the likelihood as

$$L = \prod_{i=1}^{k} \left(\prod_{l \in T_i} (1 - \alpha_i^{\exp(z_l \beta)}) \prod_{l \in R_i - T_i} \alpha_i^{\exp(z_l \beta)} \right). \qquad (18)$$

Taking for β the values $\hat{\beta}$ obtained by estimating the marginal or partial likelihood and deriving the logarithm of the above likelihood provides us with the α_i as solutions of

$$\sum_{j \in T} \frac{\exp(z_j \hat{\beta})}{1 - \hat{\alpha}_i^{\exp(z_j \hat{\beta})}} = \sum_{l \in R_i} \exp(z_l \hat{\beta}). \qquad (19)$$

If only one event occurs in t_i, the analysis gives

$$\hat{\alpha}_i = \left(1 - \frac{\exp(z_i \hat{\beta})}{\sum_{l \in R_i} \exp(z_l \hat{\beta})} \right)^{\exp z_i \hat{\beta}}; \qquad (20)$$

otherwise, an iterative solution is necessary with a recommended value α_{i0} such as:

$$\log \alpha_{i0} = \frac{-d_i}{\sum_{l \in R_i} \exp(z_l \hat{\beta})}, \qquad (21)$$

which is obtained by substituting the estimator of the following α_i in (19):

Methods of Semi-parametric Analysis

$$\hat{\alpha}_i^{\exp z_j \hat{\beta}} = \exp(\exp(z_j \hat{\beta}) \log \hat{\alpha}_i), \qquad (22)$$

which is close to

$$\hat{\alpha}_i^{\exp z_j \hat{\beta}} \simeq 1 + \exp(z_j \hat{\beta}) \log \hat{\alpha}_i. \qquad (23)$$

Let us now consider the maximization algorithms that are used.

9.3. The Newton–Raphson Algorithm

The method of estimation using the Newton–Raphson algorithm has already been presented, in Section 8.3.2. Let us only recall that the expression of the partial likelihood takes the form

$$PL(\beta) = \prod_{i=1}^{k} \frac{\exp(s_i \beta)}{\left[\sum_{l \in R_{t_i}} \exp(z_l \beta)^{d_i}\right]}. \qquad (24)$$

The estimator of the maximum likelihood $\hat{\beta}$ is therefore obtained as a solution of $U(\hat{\beta}) = 0$; that is,

$$\sum_{i=1}^{k} (s_{ji} - d_i A_{ji}(\beta)) = 0, \qquad (25)$$

where s_{ji} is the jth component of

$$s_i = \sum z(t_i)$$

and $A_{ji}(\beta)$ is given by

$$A_{ji}(\beta) = \frac{\sum_{l \in R_i} z_{jl} \exp(z_l \beta)}{\sum_{l \in R_i} \exp(z_l \beta)}. \qquad (26)$$

To use a Newton–Raphson iteration to obtain β which maximize $PL(\beta)$, we calculate

$$I_{hj}(\beta) = -\frac{d^2 \log PL(\beta)}{d\beta_h \, d\beta_j} = \sum_{i=1}^{d} d_i C_{hji}, \qquad (27)$$

where

$$C_{hji} = \frac{\sum_{l \in R_i} z_{hl} z_{jl} \exp(z_l \beta)}{\sum_{l \in R_i} \exp(z_l \beta)} - A_{hi}(\beta) A_{ji}(\beta). \tag{28}$$

In our case, the matrix of variance–covariance estimated is given by $I_{h_j}(\hat{\beta})^{-1}$ and the vector of parameters estimated is $A_{ji}(\hat{\beta})$. It follows that

$$\hat{\beta} = [I_{hj}(\hat{\beta})]^{-1} A_{ji}(\hat{\beta}). \tag{29}$$

9.4. The Choice of a Model for the Analysis of Interactions

Our interactions approach leads us to attach a primary importance to the failure time of a modifying event (or secondary event) if it occurs before the event under study, as a point of inflexion in the individuals' behaviour patterns. This enables us to modelize a single relation between the hazard rate and individual variables, including an indicator announcing the occurrence of the modifying event such as

$$h(t) = h_0(t) \exp(z_t \beta), \tag{30}$$

where z_t is a vector of characteristics that includes a coordinate z_{it} equal to 0 if the second event has not yet occured at time t and to 1 otherwise.

However, to give priority to this modifying effect, we have chosen not to consider it simply as an explanatory variable dependent on duration. Rather, we will assess the extent of this modifying effect, if any, on the various variables introduced. Thus, following Crowley and Hu (1977), we use a proportional hazard modelization which distinguishes between the formulation of the hazard rates before and after the potential modifying effect:

$$h_{..}(t \mid u) = h^*(t) \exp[z\beta_1 + H(t-u)(\beta_0 + z'\beta_2)], \tag{31}$$

where

$$H(x) = \begin{cases} 0 & \text{if } x > 0 \\ 1 & \text{if } x \leq 0 \end{cases}$$

and u is the time of the secondary event.

It is also possible to write, using the notations of Fig. 5.2 above,

$$h_{01}(t) = h^*(t) \exp(z\beta_1) \tag{32}$$

$$h_{21}(t) = h^*(t) \exp(z\beta_1 + \beta_0 + z'\beta_2), \quad (33)$$

where $h^*(t)$ is the non-parametric baseline hazard, β_1, β_0, and β_2 represents the coefficients to be estimated from variables z (introduced before the occurrence of the modifying phenomenon) and z' (introduced after the modifying phenomenon, so that they can be the same as the previous ones or others acquired on the occurrence of this phenomenon). This model implies that the different variables have a multiplicative effect on the estimated rate.

For each characteristic, we therefore obtain its main effect (before the occurrence of the secondary event), which may be affected after the secondary event has occurred by the values of parameters β_0 and β_2. This formalization of the model also implies that the non-parametric baseline hazard remains the same before and after the occurrence of the modifying event. The model we have highlighted can be implemented only by using a single commercial package (BMDP), which involves, however, major secondary programming (see Appendix). In this case, therefore, we have developed a particular package EVACOV. The following results have all been obtained using this computer medium.

9.5. Some Applications

To illustrate the possibilities of analysis offered by the chosen model, let us examine again two examples treated using data of the 'Triple Biographie' survey.

In the study of the interactions between nuptiality and a job in farming (Courgeau and Lelièvre, 1986), the men and women of the farming sector have very differentiated behaviour patterns. Semi-parametric analysis makes it possible to identify and accurately characterize the observed behaviour. The chosen model (31) is first applied to the separate analysis of the effect of each of the characteristics on nuptiality and on the exit from farming.

In the disturbing effects of marriage (respectively, of the exit from farming), we find the unilateral interaction influences: marriage keeps women in farming, whereas for men leaving farming is an incentive to marry. On the other hand, the variables that characterize the family background and the position of the individual in it have a much slighter effect on men's behaviour than on women's. This is true for marriage as well as for women's exit from farming (Table 9.1).

TABLE 9.1. Values of the parameters in the expression of the hazard rates of nuptiality and exit from farming

	Women			Men		
	Main effect β_1	Modifying phenomenon β_0	Interaction β_2	Main effect β_1	Modifying phenomenon β_0	Interaction β_2
Nuptiality						
Eldest	0.151	0.098	−0.483**	−0.171*	0.344**	0.144
Father a farmer	−0.208**	−0.124	0.067	0.182	0.383**	0.032
Exit from agr.		−0.034			0.384**	
Exit from agriculture						
No. of siblings	0.017**	0.804**	−0.002	0.003	−0.170	0.000
Eldest	−0.352**	−0.884**	0.238	−0.067	−0.161	−0.046
Father a farmer	−0.949**	−1.274**	0.658**	−0.563**	−0.483**	0.490**
Marriage		−0.806**			−0.125	

* Result significant at 10% level.
** Result significant at 5% level.

TABLE 9.2. Optimal model of exit from farming for women (values of the parameters)

Set of variables	Main effect β_1	Disturbance β_0	Interaction β_2
No. of siblings	0.012**		0.000
Eldest	−0.320**		0.296
Father a farmer	−0.928**		0.806*
Marriage		−0.228	
Female agr. worker at marriage			−1.040
Husband a farmer			−0.359**
Father-in-law a farmer			−0.126

* Result significant at 10% level.
** Result significant at 5% level.

When all variables are introduced simultaneously into the model, it is possible to identify real strategies. Let us take the example of the women presented in Table 9.2. We can tell from this table which women remain in the agricultural sector once married:

—Before marrying, the elder daughter of a farmer with only two children has a relative risk[1] equal to $\exp[0.012 + (-0.320) + (-0.928)] = 0.290$, whereas the second daughter in a family of agricultural workers with four children has a relative risk of leaving the agricultural sector equal to $\exp[4 \times (0.012)] = 1.048$. Her leaving agriculture is thus 3.85 more likely than that for the first type of woman.

—Once married, the characteristics acquired through marriage in agriculture do not favour leaving it. In fact, the women who remain are of a very particular type. They are the elders of not too numerous siblings, daughters of farmers, women married to farmers—women who 'make a career' in farming.

Let us consider, as a second example, the purchase of the first dwelling once the family is completed (Courgeau and Lelièvre, 1988 b). For the married women in the sample who are observed beyond their child-bearing life, the end of the family formation is recorded, and 66 per cent of these couples become home-owners for the first

[1] The hazard rate is calculated as the product of the baseline hazard and of the relative risk which characterize each individual, since it is calculated from the value of the individual characteristics.

TABLE 9.3. Values of the parameters in the expression of hazard rates during the child-bearing life of couples when the husband is a skilled worker

Variables	Main effect β_1	Modifying phenomenon[a] β_0	Interaction β_2
Diploma	−0.022	−0.347**	0.273*
No. of siblings	−0.057**	−0.164	−0.003
Diploma held by spouse		−0.375**	0.316**
No. of siblings to the spouse		0.095	−0.076**
Completed fertility		0.304*	−0.206**

[a] Here the birth of the last child.
* Result significant at 10% level.
** Result significant at 5% level.

TABLE 9.4. Values of the parameters in the expression of hazard of purchase once the family is formed (couples where the husband is a skilled worker)

Variables	Main effect β_1
Diploma	0.538**
No. of siblings	0.005
Diploma held by spouse	0.594**
No. of siblings to the spouse	0.013
Completed fertility	−0.002

** Result significant at 5% level.

time after this event (with regard to cohorts born between 1911 and 1935). We have therefore calculated the effects of the variables considered separately for five occupational groups of spouses.

The purchase during the fertility period is always less probable for couples where the husband is a skilled worker (Tables 9.3 and 9.4), coming from a large family, or having many children. Once the family is formed, the purely family constraints lose ground until their role is no longer significant. Moreover, the possession of diplomas, which gives indication as to the career of cohorts' spouses, appears distinctly favourable, whereas these do not have a primary role on precocious purchases, which instead tend to be determined by a rural origin.

TABLE 9.5. Estimated parameters for the entry into local authority accommodation in England subsequent to marriage in 1961–5, according to the number of prior births (time-dependent variable)

Variable	Main effect
First birth	0.720**
Second birth	0.142
Third birth	0.863***
Log-likelihood	– 1669.56

***Significance level: $p < 0.001$.
Source: Murphy (1984)

It is also possible to introduce time-dependent variables in semi-parametric models. Thus, M. Murphy (1984), when analysing the entry into a local authority accommodation in England subsequent to the individuals' marriages, introduced this kind of variable for successive births. For example, as long as the first birth has not occurred, this variable is nil. It becomes equal to 1 after the first birth. Table 9.5 reports these results and demonstrates the truly significant effect of the first and third births: the probability of having local authority accommodation is multiplied by 2 after the first birth and by 2.37 after the third one.

These examples show how analysis makes it possible to be more precise and thus to explain the behaviour observed. Some hypotheses put forward in more qualitative studies can thus be invalidated or confirmed. Others are suggested, which then must be examined at greater depth through other disciplines (qualitative sociology or psychology, for example). These very elaborate analyses provide material for pluri-disciplinary interaction. The results obtained require a knowledge of other disciplines, or call on them to join in an exploration of an area of analysis.

9.6. Conclusion

This analysis is more flexible than a purely parametric modelization. Therefore it represents an interesting alternative solution, when the objective is to measure the influence of individual characteristics on the hazard rates estimated.

Moreover, if we do not want to take explicit account of the unobserved heterogeneity of the individual behaviour described, the non-parametric element 'takes over' this variance without imposing anything on its distribution.[2] The current software (BMDP, RATE, GLIM) do not always take explicit account of this option, but with a little more programming, it should be possible. What is not always possible is to take account of time-dependent characteristics.

[2] During the IUSSP seminar on event history analysis (Paris, March 1988), among the six contributions that were requested in order to compare the methods, four teams chose a semi-parametric model of analysis.

10

Conclusion

In the Introduction to this book we presented its two main objectives: on the one hand, to develop methods that make it possible to analyse the interactions between demographic phenomena; on the other hand, to examine and deal with the heterogeneity of observed populations. While both these problems had already been raised some time ago (Henry, 1959; Pressat, 1966), no satisfactory solution has thus far been found.

Most often, the option was taken to eliminate the modifying effect that one phenomenon had on another, without analysing the more complex interactions between phenomena. Similarly, populations were considered to be homogeneous or, at best, were broken down on the basis of simple criteria into sub-populations which were dealt with separately.

The collection of individual biographies, which are obtained increasingly often through retrospective surveys, has enabled us to come up with new answers to both these problems. Throughout the present work the methods we have presented analyse the interactions between demographic, social, and economic phenomena while at the same time bringing into play the heterogeneous nature of the observed populations. These methods have now become a valuable tool for longitudinal analysis, as is shown by their increasing use by demographers and the many results they have helped bring to light.

In our Conclusion we shall make a synthesis of the solutions brought by these methods to the two problems stated. We shall also attempt to identify new areas of research which are opened up by these methods.

10.1. Analysis of Interactions between Phenomena

The basis of this first approach is the collection of the various events that influence an individual's existence. This means that the approach goes beyond that of the analysis of phenomena that are purely demographic, since the events may be of a very diverse nature.

These may include, for instance, events of a purely physical nature which may affect the existence of a few or even millions of individuals. Such events are generally recorded in sources other than the individual biography and therefore can easily be added to the latter. Earthquakes, volcanic eruptions, severe winters, etc., are all events independent of the societies they affect, yet they will influence and alter the life-courses of the individuals subject to their occurrence.

Biological events may also come into play: puberty, menopause, giving birth, etc. In this case, the occurrence of such an event may well be different depending on the society in which the individual lives. But more important, from another standpoint, we can study the extent to which these events can alter the existence of those who experience them.

Other events of a social, economic, and political nature will also have important consequences on the life of an individual. Some of these events may depend on the individual's past life. Thus, the fact that an individual becomes an engineer is related to the different diplomas that he has obtained during school and university education. Moreover, the fact of becoming an engineer will have an effect on future events in his life. In the same way, an event of a political character, such as participating in a strike action, may influence an individual's career.

Finally, we meet more complex events of a psychological nature. The beginning of a friendship, for instance, or an emotional attachment to a place may modify an individual's future attitudes. Most often, it will be difficult to enter these events into a demographic survey. But they may be highly important and should not be neglected.

All these events can be situated with varying precision within an individual's life-course. For the demographer, they are the equivalent of what particles represent for the physicist. Thus, the number of events to be considered is not fixed once and for all, but varies from one period of time to another, from one society to another. In the same way, the study of the interaction between events can focus on two or more types of events and can eliminate the modifying effect of other events. Once the attention is focused on the successive dates of occurrence of events, it becomes possible to theorize the analysis to be carried out.

10.1.1. Theorization

At this point, we are not going to repeat in detail the formalization which we presented in the introduction to Chapter 2. We shall, however, be more precise about the hypotheses underlying it.

If the interactions between phenomena are to be analysed, we must first postulate that each event has an underlying probability of occurrence and that this probability changes with time and is modified when other phenomena intervene prior to its occurrence.

The observation of a single biography does not permit these probabilities to be estimated, as only one finished process/reference is available. When, however, a sample of individuals is observed, it is possible to estimate these probabilities.

If the probabilities are to be meaningful, other hypotheses are required. An important point is that the study must be done on a sub-population that is homogeneous enough to allow the interactions studied to appear clearly. For this reason, we isolated a sub-population of unmarried people who began their working life in the agricultural sector, in order to bring out the links between marriage and a departure from agriculture. The next step, of course, is to try and see if there are other origins of heterogeneity in the sub-population chosen. This is discussed in the section on heterogeneity.

Observing a homogeneous population also means observing the members of a well-defined generation or cohort. The hypothesis that behaviour from one generation to another is stable cannot be taken for granted, and is unlikely to be proved. What is proposed here, therefore, is a study of the *evolution* of behaviour from one generation to the next.

The use of retrospective survey data to highlight this evolution in behaviour results in new hypotheses being put forward. With this method of observation, it is not the entire existence of an individual that is recorded, but only those events that occur prior to the survey. We thus obtain right-censored biographies, and in order to reach a correct estimate of the probabilities, we adopt the underlying hypothesis that the individuals who have not experienced the studied events were exposed to the same probability of occurrence of these events over the observed period as those who actually experienced them. Again, what is being studied here is a condition of homogeneity.

It can be seen, therefore, that the analysis of interaction between phenomena is perfectly justified when a sufficiently homogeneous sub-population is analysed over a relatively long period of time. It must be noted that this hypothesis, related in the present case to a sub-population, is in fact identical to that used in traditional demography when a country's entire population is being worked on. Breaking this larger population down into sub-populations can only contribute an even greater degree of homogeneity and from a methodological point of view constitutes a generalization of methods used in traditional demographic analysis.

10.1.2. Non-parametric models

After postulating these hypotheses, we then showed that non-parametric models make it possible to estimate, without further hypothesis being necessary, the probabilities of transition from one state to another.

Obviously, the more these models can make use of different states, the more precise they will be. The univariate model used in traditional demography would seem quite inappropriate in this case, since it mixes the behaviour patterns of individuals in very different situations regarding other phenomena which interfere with the phenomena under study. It was, however, the main model used until very recently.

It is now helpful to use the bivariate model, which permits a very detailed analysis of the interactions between two phenomena. These interactions have been seen to be very complex and to allow a better understanding of human behaviour.

We have brought to light various kinds of dependency. The lowest level of dependency is when the first phenomenon is independent of the second, and the second is independent of the first. This *total independence* between several phenomena has so far never been observed in the analysis of the 'Triple Biographie' data. It seems to be rare in human populations, a fact which shows that the various sociological, economic, political, etc., aspects of human phenomena are closely interrelated.

A *unilateral dependence* is considerably more interesting to observe. This means that the occurrence of one of the events modifies the probability of the second event occurring; but, conversely, the occurrence of the second event in no way affects the probability of occurrence of the first event. This type of unilateral dependence very often appeared in the analyses that we carried out on data from the 'Triple Biographie' survey. We were able to discern that a departure from the agricultural sector has no influence on marriage as far as women are concerned; yet, once married, they tend to stay in that occupational sector much more than single men do. As far as men are concerned, we discovered a unilateral dependence contrary to that for women: their chances of getting married double when they leave the agricultural sector (Courgeau and Lelièvre, 1986).

Finally, a *reciprocal dependence* between two phenomena was also often observed. Each of the two phenomena, when they occur, modify the probabilities of occurrence for the other. For instance, if the probability of migration to an urban area decreases after each birth, the likelihood of the birth of a second or subsequent child decreases following a migration to a metropolitan area (Courgeau, 1987c).

The dependencies can be observed only at certain ages, or over a given period following the modifying event. Moreover, the influence can be reversed. For example, when a study is made of the links between fertility and female activity, it can be seen that those women who are not active at the time of their marriage and at the birth of the first child have a different fertility according to their age: while at a young age those women who have started to work again are as fertile as those who remain inactive, after the age of 30 resuming an active life constitutes a brake to further child-bearing (Lelièvre, 1987 *a*).

Finally, other more complex levels of interpretation can be demonstrated. Thus, in the study of the interaction between fertility and migration between metropolitan and non-metropolitan areas, the birth pattern for the second or subsequent child was found to be considerably modified. The question, however, is whether this is to do with an adaptation or a selective behaviour pattern: adaptation if what modified the fertility of the migrant women was indeed the migration; selection if in the area of departure a behaviour pattern is observed which is different for future migrant women and for sedentary women. Again, the methods presented in the present work make it possible to test the differences between the future migrant women and the ultimately sedentary women in the initial population. It can be shown that, in France, the *selection* hypothesis is confirmed for women who migrate to a metropolitan area. These future migrants do in fact already have a low level of fertility compared with sedentary women of non-metropolitan areas. This fertility is the same level as that of women who have already migrated (Courgeau, 1987c).

Thus, it can be seen that there exists an *a priori dependence* of fertility on future migration, which operates this selection within the initial population. In addition, an increase in the level of fertility can be observed for those women who move to less urbanized areas. Using a method of investigation that is identical to the previous one, it is possible to show that the fertility behaviour pattern of those women who move away from metropolitan areas undergoes a process of adaptation; their previous fertility behaviour patterns are in no way different from women who live and who remain in metropolitan areas.

These bivariate methods may be completed by the trivariate or multivariate methods also presented in this work. Although the estimation of these models raises no complex problems at a theoretical level, in practice, the number of individuals observed does not generally make it possible to take a non-parametric analysis much further in this direction. In effect, the number of hazard rates to be estimated does rapidly increase as the number of possible situations increases. The population at risk is thus very soon insufficiently numerous to allow any precise estimation.

A similar problem occurs when one tries to introduce the heterogeneity of observed populations. If non-parametric methods are still to be used, the studied population needs to be broken down into sub-populations that are sufficiently homogeneous regarding the different characteristics that one wishes to focus on: social origin, number of brothers and sisters, number of births, diplomas, etc. Here again, the number of individuals within the sub-populations observed will be reduced when the number of characteristics under consideration increases. It will thus be impossible to finalize the analysis when these groups become evanescent.

Other methods must therefore be used to analyse this heterogeneity.

10.2. Dealing with heterogeneity in populations

The events that arise during a person's life are not the only factors that contribute to his individual life-course. Many other characteristics, possessed from birth or acquired during childhood, are important elements which have different influences during his life.

Family origins, for instance, play a role. Having a father who is an agricultural worker, an industrial worker, or a senior executive will already place an individual in a given milieu and this will obviously affect his future career. Similarly, being the youngest or eldest child, being an only child or having many siblings, being a girl or a boy, being born into a rural or an urban environment, etc., all constitute characteristics that will affect an individual's future life.

This heterogeneity can, moreover, be generalized if one is studying an individual's life-course from a starting point in time other than birth. Thus, when successive migrations are being studied, this starting point can correspond to moving into a new dwelling. One could then associate with the duration of stay in this new dwelling all the various characteristics concerning the individual that are present at the beginning of the stay: age, marital status, number of children, job, etc. (Courgeau, 1985 *b*, *d*). Of course, family origins may also have their place among all these characteristics.

To deal with this heterogeneity, the parametric methods have proved very useful.

10.2.1. Parametric models

These models constitute the generalization to the analysis of durations of stay of regression models used, for example, in econometry. They require more restrictive hypotheses than those for the non-parametric models presented above. In fact, not only the duration of stay needs to be modelized, but also the effect of various characteristics on the occurrence of the event under study.

In Chapter 8 we presented a wide variety of parametric models which enabled us to express the distribution of durations of stay using a small number of parameters. These models provide hazard rates that either uniformly increase, decrease, or remain constant over time, as well as those hazard rates that attain a maximum point before decreasing. All such types of distribution can be observed in demography. Moreover, multimodal distributions can be constructed by combining some of the previous distributions.

When several of these distributions can be adapted with the same degree of accuracy to an observed distribution, it is preferable to choose that which has a hazard rate and a survivor function with forms that are simple and explicit. This makes it much easier to estimate the parameters on the basis of data that are partially censored.

Once the duration of stay has been modelized, the effect of the various individual characteristics on the observed hazard rates must be introduced. In order to do so, we have presented two main types of model.

The proportional hazard model supposes that the hazard rates of individuals with a given characteristic are proportional to the rate of individuals who do not have the characteristic, and that the factor of proportionality is the same for all durations. Obviously this very strong hypothesis must be tested with regard to all the characteristics under consideration. If the hypothesis is not verified, the populations then need to be broken down into sub-populations, for which two proportional hazard models are estimated for all the other characteristics. The possibility that other models may be better adapted to these data should also be examined. An example of this is the analysis we carried out of moves to new dwellings in relation to some thirty characteristics of individuals at the beginning of their stay. Owing to the fact the individuals' sex resulted in hazard rates that could not be considered as proportional, the sampling had to be broken down into two sub-populations differentiating between men and women (Courgeau, 1985 *b*, *d*).

The second accelerated failure time model supposes that the characteristics have a direct influence on the time factor of an individual's life. This means that individuals with a given characteristic will experience the studied event in terms either more accelerated or slower than those individuals who do not have the characteristic.

Both these models seem to be the most adapted to human behaviour patterns. They have been successfully used in a large number of studies. Of course, many other types of models can be used, and it may be that in future some of these may prove more complete than the proportional hazard model or the accelerated failure time model. A linear model has occasionally been used, which adds a positive or negative constant to the hazard rate whenever an individual has a given characteristic. Unlike the proportional hazard model, an individual may in this case have a negative estimated hazard rate. This eventuality therefore indicates a disadvantage of using the linear model.

In all cases, these parametric models may introduce only characteristics that are observed during the survey. It could be imagined, however, that other characteristics which are more difficult to observe or measure during a survey or which, in the researcher's opinion, do not influence the event studied do in fact affect hazard rates to an extent that cannot be neglected. This unobserved heterogeneity may

Conclusion

affect the parameters used to measure the effect of the observed characteristics. This is the reason for the introduction of a parametric or even non-parametric distribution of this unobserved heterogeneity, which has a multiplicative effect on the duration-of-stay distribution. It can be shown that, under this condition, it is possible to estimate new parameter values which correspond to the observed characteristics and which take into account this unobserved heterogeneity. It seems that, according to the supposed distribution of this unobserved heterogeneity (Heckman and Singer, 1984), or even to the parametric distribution adopted to estimate the effect of characteristics (Trussell and Richards, 1985), the estimated parameters may vary enormously and may even indicate contrary results.

These results mean that the semi-parametric approach we are now developing tends to be favoured. This approach has given more precise results concerning the effect of unobserved heterogeneity on the estimation of parameters. It also makes it possible to take simultaneously into account both the heterogeneity of populations and the interactions between phenomena.

10.2.2. Semi-parametric models

These models are much more flexible than the preceding ones, as here the effect of the duration of stay on the hazard rate is no longer in the form of a parametric model. On the other hand, they do introduce a similar effect of the various characteristics observed. As a result, we obtain in much the same way proportional hazard models, accelerated failure time models, or any other type of link between the duration of stay and the observed characteristics.

In Chapter 9 we showed how to estimate the effect of characteristics using the partial likelihood method, and the non-parametric effect of the duration of stay on the hazard rate.

These models have a certain number of advantages over the preceding models. Firstly, it is possible to study on the theoretical level how omitting characteristics in a model of this kind affected the estimated parameters of the observed characteristics (Bretagnolle and Huber-Carol, 1985). It was shown that this omission had no effect on the sign of the estimated parameters, but that it did result in a reduction of the absolute values of these parameters. This means that, if the effect of a characteristic had appeared as important when other independent ones were omitted, introducing them into the semi-parametric model will only reinforce the effect of the first characteristic. On the other hand, some characteristics that seemingly had no significant effect may acquire a pronounced significance when characteristics initially unobserved are introduced.

These results are very important, as they allow us to be sure about the meaning of the observed effects, even though we do not know whether all the characteristics affecting the duration of stay have been introduced into the model.

Secondly, this model makes it easy to introduce simultaneously interactions between demographic phenomena and the heterogeneity of observed populations. Thus, in the bivariate type model different underlying hazard rates can be introduced for different types of events. One can even introduce hazard rates that have related proportionalities, once this has been verified by non- parametric analysis. It was this last solution that we chose in order to analyse the relation between nuptiality, the exit from agriculture, and various individual characteristics (Courgeau and Lelièvre, 1986).

There exists another possibility of introducing more complex interactions, where characteristics are modified according to the duration of stay. Here, it should be noted that the use of a purely parametric model makes it possible to introduce characteristics that depend on elapsed time in a similar way.

10.3. New lines of research

Collecting and analysing life histories according to the methods presented in the present work opens up a wide field of research which has only just begun to be explored. Of course, this field is to be situated within a much wider context which includes all the social sciences. Among anthropologists, psychologists, sociologists, etc., there are many researchers using event history analyses, and the hypotheses and methods they use vary greatly from one discipline to another. Even the basic material—the event history—is often collected in such different ways that one researcher would find it very difficult to use event histories compiled by a colleague working in another discipline.

Despite these differences, it would seem important that each of the individual sciences should open itself up to the others. One of the aims of the present work is to show more clearly that demography has an original contribution to make in this respect.

Firstly, we have repeatedly referred to the fact that for a demographer it is very difficult, if not impossible, to collect and treat both the psychological characteristics of an individual and the sociological characteristics of a group. This brings in the element of an unobserved heterogeneity, which as we have seen poses a great many problems. A common line of research on this subject would seem very desirable and would doubtless prove extremely enriching for the social sciences. A second line of research would involve reflecting jointly on the theoretical bases of our methodologies. It would be useful to try and see whether there exists a common theoretical core within the different branches of social science which make use of event history analysis, and to try and identify its limits. A more precise definition could be attempted of exactly what each branch is exploring and why harmonizing the different viewpoints on the same phenomenon appears to be a difficult matter. After all, the data of a research survey, whether carried out by an anthropologist, a psychologist, a sociologist, or some other social scientist, are provided by the same individual, though of course, a different viewpoint and different objectives are involved. Once they have collected their images, they will seem to be working on different subject matters. In this case, it is important to analyse the links that exist between these subjects.

It is also of interest to see whether the methods of event history analysis can be generalized to more complex social structures such as the history of a family, an enterprise, a nation. And is it possible to use the same methods of analysis as for an individual biography? Similarities do exist between histories: individual events can greatly affect not only a family but also an enterprise. The unit of analysis, however, is much more complex, as a family or enterprise is made up of a great many individuals whose relationships and changes in relationships are part of the events to be analysed. This constitutes yet another important area to be explored.

Finally, these conclusions show that the methods of analysis we have developed throughout this work go beyond the individual towards a deeper understanding of human societies. Certainly, these individual biographies contain a wealth of information on the society that shapes but is also shaped by them. It has thus been our aim to exploit the information regarding an individual in the best possible way so as to throw new light on human behaviour patterns.

Appendix

A1. Basic Formulas

Survivor function

- Continuous time

$$S(t) = P(T \geq t), \qquad S(0) = 1$$

$$S(t) = \int_t^\infty f(s)\,ds$$

$$S(t) = \exp\left(-\int_0^t h(s)\,ds\right) = \exp(-H(t))$$

- Discrete time

$$S(t) = \sum_{i \mid t_i > t} P(T = t_i)$$

$$S(t) = \prod_{t_i < t} (1 - h_i)$$

Probability density function

- Continuous time

$$f(t) = \lim_{\Delta t \to 0} \frac{P(t \leq T < t + \Delta t)}{\Delta t}$$

$$f(t) = -\frac{dS(t)}{dt} = -S'(t)$$

$$f(t) = h(t) \exp\left(-\int_0^t h(s)\,ds\right) = h(t) \exp(-H(t))$$

- Discrete time

$$f(t_i) = P(T = t_i)$$

Appendix

Hazard rates (instantaneous failure rates), conditional probability density

- Continuous time

$$h(t) = \lim_{\Delta t \to 0} \frac{P(T < t + \Delta t \mid T \geq t)}{\Delta t} = \lim_{\Delta t \to 0} \frac{P(t \leq T < t + \Delta t)}{\Delta t\, P(T \geq t)}$$

$$h(t) = \frac{f(t)}{S(t)} = -\frac{dS(t)/dt}{S(t)} = -\frac{d \log S(t)}{dt}$$

- Discrete time

$$h(t) = \sum_i h_i\, \delta(t - t_i)$$

where $\delta(0) = 1$
$\delta(x) = 0, \quad \forall\, x \neq 0$
$h_i = f(t_i)/[f(t_i) + f(t_{i+1}) + \ldots]$

The intensity of the process can also be estimated:

$$H(t) = \sum_{i \mid t_i < t} \log(1 - h_i)$$

If h_i are 'small',

$$H(t) \simeq \sum_{i \mid t_i < t} h_i$$

Bivariate case

Let T_1 and T_2 be the duration at which two specific events occur:

- Hazard rates

$$h_{0i}(t) = \lim_{\Delta t \to 0} \frac{P(T_i < t + \Delta t \mid T_i \geq t,\, T_j \geq t)}{\Delta t}$$

$$h_{ji}(t \mid u) = \lim_{\Delta t \to 0} \frac{P(T_i < t + \Delta t \mid T_j = u,\, T_i \geq t)}{\Delta t}, \quad u \leq t$$

- Probability density function

$$f_i(t_i) = h_{0i}(t_i) \exp - \int_0^{t_i} \left(h_{0i}(u) + h_{0j}(u) \right) du$$

$$f_j(t_j \mid t_i) = h_{ij}(t_j \mid t_i) \exp \left(- \int_{t_i}^{t_i} h_{ij}(u \mid t_i) du \right), \quad t_i \leq t_j$$

Hence the joint density:

$$f(t_i, t_j) = h_{0i}(t_i)\, h_{ij}(t_j \mid t_i)\, \exp\left[-\int_0^{t_i}(h_{0i}(u) + h_{0j}(u))du - \int_{t_i}^{t_j} h_{ij}(u \mid t_i)du\right], \quad t_i \leq t_j.$$

A2. Existing Packages

No systematic review is here undertaken of all the existing packages available for event history analysis.

Here we shall only refer to five packages available worldwide—LIFETEST (SAS), LIFEREG (SAS), GLIM (NAG), PL1 (BMDP) and PL2 (BMDP)—and to two packages written in the INED by the authors.

Two packages for non-parametric analysis

LIFETEST (SAS) and PL1 (BMDP) offer approximately the same facilities. Section 4.3 shows results obtained with LIFETEST. Let us give some of the instructions:

PROC LIFETEST	calls the procedure
TIME	names the sojourn variables and the censoring indexes
STRATA	breaks down the sample into sub-groups
TEST	calls statistical tests within each strata
FREQ	indicates the distribution of each variable
BY	can also be used to sort the sample but no test of the homogeneity of each group can then be undertaken

In Chapter 4, the following instructions were given:

DATA don1; SET survie.grp;
PROC SORT DATA = don1; BY sex groupof;
PROC LIFETEST;
TIME dur1*cens1;
STRATA groupof;
By sex;

---

[1] Among others: LOGLIN, documented in LOGLIN 1.9 User's Guide (1965), written by Olivier and Neff (Harvard University Public Health Science Computer Facility), is widely used by J. Hoem; SURVREG, written by Preston and Clarkson in 1983, estimates parametric proportional hazard rate models; PHGLM (1983) is a package written by users but not documented by SAS.

Appendix

The procedure for PL1 is equally simple to use, as no specific form is required for the data.

The analysis of interaction between two events: ROOT (INED)

This package was written at the INED in May 1987 by Eva Lelièvre for the non-parametric analysis of interactions between two events, as no such package was available. It gives an estimation of hazard rates and provides different tests. It is very flexible, allowing a modification of the time-scale and providing different options to treat the simultaneities. It was written in FORTRAN and is documented in French. Since October 1987, it is distributed by the Association Nationale du Logiciel.

Parametric and semiparametric packages

- GLIM (McCullagh and Nelder 1983) is designed for the estimation of log-linear models:

$$h(t, z) = h_0(t) \exp(\beta z).$$

What is explored is

$$p_i(z) = P(T < t_i \mid T > t_i - 1; z).$$

The models in GLIM specify the form of $p_i(z)$ through a known function (logit, normal, complementary log-log) described for the 'link function'.

Diamond, McDonald, and Shah (1986) describe the way to estimate semi-parametric models with GLIM.

- LIFEREG (SAS) offers three choices for the 'link function': exponential, Weibull, and gamma.
- PL2 (BMDP) is the only package able to estimate time-dependent covariates.
- RATE, developed by N. Tuma, can estimate parametric and semi-parametric models.
- EVACOV (INED, 1986) was developed from the program proposed by Chang, Wang, and McIntosh (Kalbfleisch and Prentice, 1980). It allows a semi-parametric estimation of the interactions between events. Eight fixed covariates as well as 14 time-dependent covariates can be incorporated in the model, the latter being modified only at the occurrence of the modifying event. Applications in Chapter 9 were obtained using this package.

Bibliography

AALEN, O. (1977). 'Weak convergence of stochastic integrals related to counting processes'. *Zeitschrift für Wahrscheinlichkeitstheorie*, 38: 261–77.
—— (1978). 'Nonparametric inference for a family of counting processes'. *Annals of Statistics*, 6: 701–26.
—— (1982). 'Practical applications of the nonparametric statistical theory for counting processes'. Statistical Research Report no. 2, Institute of Mathematics, University of Oslo.
—— and HOEM, J. (1978). 'Random time changes for multivariate counting processes'. *Scandinavian Journal of Statistics*, no. 2: 81–101.
—— and JOHANSEN, S. (1978). 'An empirical transition matrix for non-homogeneous Markov chains based on censored information'. *Scandinavian Journal of Statistics*, 5: 141–50.
—— O., BORGAN, N., KEIDING, and THORMAN, J. (1980). 'Interaction between life history events: nonparametric analysis for prospective and retrospective data in the presence of censoring'. *Scandinavian Journal of Statistics*, 7: 161–71.
ALLISON, P. D. (1980). *Event History Analysis: Regression for Longitudinal Event Data*. Beverly Hills: Sage.
—— (1982). 'Discrete-time methods for the analysis of event histories'. In *Sociological Methodology*, ed. S. Leinhardt. San Francisco: Jossey-Bass, pp. 61–98.
ANDERSEN, P. (1980). 'Testing goodness-of-fit of Cox's regression and life model'. Research Report no. 80/1, Danish Statistical Research Unit.
—— (1981a). 'Comparing survivals via hazard ratio estimates'. Research Report no. 81/7, Danish Statistical Research Unit.
—— (1981b). 'Measuring and evaluating prognosis using the proportional hazards model'. Research Report no. 81/8, Danish Statistical Research Unit.
—— (1981c). 'On the application of the theory of counting processes in the statistical analysis of censored survival data'. Research Report no. 81/10, Danish Statistical Research Unit.
—— and BORGAN, O. (1985). 'Counting processes for life history data: a review'. *Scandinavian Journal of Statistics*, 12: 97–158.
—— and GILL, R. (1981). 'Cox's regression model for counting processes: a large sample study'. Research Report no. 81/6, Danish Statistical Research Unit.
—— BORGAN, O., GILL, R., and KEIDING, N. (1982). 'Linear nonparametric

tests for comparison of counting processes, with applications to censored survival data'. *International Statistical Review*, 50.
ANDRESS, H. J. (1983). 'The first 10 years of a working career: an illustration of event-history analysis with West German mobility data'. *Computational Statistics and Data Analysis*, no. 1: 111–35.
ANNALES DE VAUCRESSON, (1987). *Histoires de vie, histoires de famille, trajectoires sociales*. Vaucresson: CRIV.
ARJAS, E. and KANGAS, P. (1992). 'Discrete-time method for longitudinal analysis in demography: a comparative study of the data on third births in Sweden'. In Trussel *et al.* (1992).
—— and VENZON, D. (1988). 'A test for discriminating between additive and multiplicative relative risk in survival analysis'. *Applied Statistics*, no. 37: 1–11.
BACK, K. W. (ed.) (1980). *Life Course: Integrative Theories and Exemplary Populations*. American Association for the Advancement of Science, Selected Symposium no. 41.
BERTAUX, D. (1980). 'L'approche biographique: sa validité méthodologique, ses potentialités'. *Cahiers Internationaux de Sociologie*, 69: 197–224.
BLOSSFELD, H. P., HAMERLE, A., and MAYER, K. U. (1986). *Ereignisanalyse: Statistische Theorie und Anwendung in den wirtschafts—und socialwissenschaften*. Frankfurt: Campus Verlag.
BLUMEN, I., MARVIN, K., and MCCARTHY, P. J. (1955). *The Industrial Mobility of Labour as a Probability Process*. Cornell Studies of Industrial Labour Relations, vi. Ithaca, NY: Cornell University Press.
BOCQUIER, Ph. (1987). 'Retours dans le pays d'origine des immigrants en Suède'. *Notes et Documents, Population*, 42: 544–8.
BOURDIEU, P. (1986). 'L'illusion biographique'. *Actes de la Recherche en Sciences Sociales*, 62–3: 69–72.
BRESLOW, N. E., LUBIN, J. H., MAREK, P., and LANGHOLZ, B. (1983). 'Multiplicative models and cohort analysis'. *Journal of the American Statistical Association*, 78: 1–12.
BRETAGNOLE J. and HUBER-CAROL, C. (1985). 'Effet de l'omission de covariables dans le modèle de Cox'. *statistiques des processus en milieu médical*, Séminaire 85, v. Paris: Huber, Lelouch, Prieur.
—— (1988). 'Effects of omitting covariates in Cox's model for survival data'. *Scandinavian Journal of Statistics*, 15: 125–38.
BRILLINGER, D. (1986). 'The natural variability of vital rates and associated statistics (a Biometrics-invited paper with discussion)'. *Biometrics*, 42: 693–734.
BROUARD, N. (1980). 'Espérance de vie active, reprise d'activité féminine: un modèle'. *Revue Economique*, 31: 1260–87.
BUCKLEY, J. D. (1984). 'Additive and multiplicative models for relative survival rates', *Biometrics*, 40: 51–62.

CAMBOIS, M. A. (1987). 'Rapport sur la mobilité professionnelle'. Evidence presented to the Inquiry on Career Mobility in Survey 3B. Paris: INED.
CHANG, M. N., and YANG, G. L. (1987). 'Strong consistency of a nonparametric estimation of the survival function with doubly censored data'. *Annals of Statistics*, 15: 536–1, 547.
CHIANG, C. (1968). *Introduction to Stochastic Processes in Biostatistics*. New York: John Wiley.
COALE, A. and MCNIEL, D. (1972). 'The distribution by age of the frequency of first marriage in a female cohort'. *Journal of the American Statistical Association*, 67: 27–52.
COLEMAN, J. S. (1981). *Longitudinal Data Analysis*. New York: Basic Books.
COLLOMB, Ph. (1987). *La mort de l'orme séculaire*, and *crise agricole et migration dans l'Ouest audois des années cinquante*. Travaux et Documents, nos. 105 and 106. Paris: INED/PUF.
COURGEAU, D. (1973). 'Migrants et migrations'. *Population*, 28: 92–129.
—— (1980). *Analyse quantitative des migrations humaines*. Paris: Masson.
—— (1984a). 'Relations entre cycle de vie et migrations'. *Population*, 39: 483–513
—— (1984b). 'Analysis of the French migration, family and occupation history survey'. *Materialen Bevolkerungwissenschaft*, BIB, 38: 86–102.
—— (1985a). 'Effet de déclarations erronées sur une analyse de données migratoires'. In *Migrations internes*. Chaire Quetelet '83. Louvain-la-Neuve: Jezierzki.
—— (1985b). 'Interaction between spacial mobility, family and career lifecycle: a French survey'. *European Sociological Review*, 1: 139–62.
—— (1985c). 'Bases théoriques et modèles pour une enquête sur la biographie familiale, professionnelle et migratoire'. *Espace, Population, Société*, 1: 240–7.
—— (1985d). 'Changements de logement, changements de département et cycle de vie'. *L'Espace Géographique*, 4: 289–306.
—— (1987a). 'Pour une approche statistique des histoires de vie'. *Annales de Vaucresson*, 26: 25–36.
—— (1987b). 'L'analyse des enquêtes rétrospectives'. In *L'explication en sciences sociales: la recherche des causes en démographie*. Chaire Quetelet '87. Louvain-la-Neuve: Jezierzki.
—— (1987c). 'Constitution de la famille et urbanisation', *Population*, 42: 57–82.
—— (1988). *Méthodes de mesure de la mobilité spatiale: migrations internes, mobilité temporaire, navettes*. Paris: Editions de l'INED.
—— (1990). 'Migration, family, and career: a life course approach'. In *Life-Span Development and Behavior*, ed. P. B. Baltes, D. L. Featherman and R. M. Lerner. Hillsdale, NJ: Lawrence Erlbaum Associates.

—— (1991). 'Analyse de données biographiques erronées'. *Population*, 46: 89–104.
—— and LELIÈVRE, E. (1986). 'Nuptialité et agriculture', *Population*, 41: 303–26.
—— (1988). 'Estimation of transition rates in dynamic household models'. In *Modelling Household Formation and Dissolution*, ed. N. Keilman, A. Kuijsten, and A. Vossen. Oxford: Clarendon Press, pp. 160–76.
—— (1992). 'Interrelation between first home ownership, constitution of the family and professional occupation'. In Trussel *et al.* (1992).
—— and WAGNER, M. (1986). 'Leaving home and marriage in France and Germany'. Unpublished paper.
COX, D. R. (1972). 'Regression models and life tables (with discussion)'. *Journal of the Royal Statistical Society*, B34: 187–220.
—— (1975). 'Partial likelihood'. *Biometrika*, 62: 269–76.
—— and HINKLEY, D. V. (1974). *Theoretical Statistics*. New York and London: Chapman and Hall.
—— and OAKES, D. (1984). Analysis of Survival Data. London: Chapman and Hall.
CROWLEY, J. (1974). 'Asymptotic normality of a new nonparametric statistic for use in organ transplant studies'. *Journal of the American Statistical Society*, 69: 1006–11.
—— and HU, M. (1977). 'Covariance analysis of heart transplant data'. *Journal of the American Statistical Association*, 72: 27–36
DAVIES, R. B. (1984*a*). 'A generalised beta-logistic model for longitudinal data with an application to residential mobility'. *Environment and Planning* A, 16: 1375–86.
—— (1984*b*). 'Calibrating longitudinal models of residential mobility and migration', *Regional Science and Urban Economics*, 14: 231–47.
—— and CROUCHLEY, R. (1985*a*). 'Longitudinal versus cross-sectional methods for behavioural research: a first-round knockout'. *Environment and Planning* A, 17: 1315–29.
—— —— (1985*b*). 'A panel study of life cycle effects in residential mobility'. *Geographical Analysis*, 17: 199–216.
DEROO, M., and DUSSAIX, A. M. (1980). *Pratique et analyse des enquêtes par sondage*. Paris: PUF.
DIAMOND, I., MCDONALD, J., and SHAH, I. (1986). 'Proportional hazards models for current status data: application to the study of differentials in age at weaning in Pakistan'. *Demography*, 23: 607–20.
DUCHENE, J. (1985). 'Un test de fiabilité des enquêtes rétrospectives Biographie Familiale, Professionnelle et Migratoire'. In *Migration internes*. Chaire Quetelet '83. Louvain-la-Neuve: Jezierzki.
ELAND-JOHNSON, R. C., and JOHNSON, N. L. (1980). *Survival Models and Data Analysis*. New York: John Wiley.

ELDER, G. (1978). 'Approaches to social changes and the family turning points'. *American Journal of Sociology*, 84, Special Number: 1–39.

FELLER, W. (1968). *An Introduction to Probability Theory and its Applications*, 2 vols. New York: John Wiley.

FINKELSTEIN, D. M. (1986). 'A proportional hazard model for interval-censored failure time data'. *Biometrics*, 42: 845–54.

FONER, A., and KERTZER, D. (1978). 'Transition over the life course: lessons from age-set societies'. *American Journal of Sociology*, 83: 1081–1105.

FOUGERE, D. and TAHAR, G. (1987). 'Participation au marché du travail et nuptialité: étude des interdépendances au sein d'une cohorte'. Note no. 60 (87–09), U A 921 du CNRS.

GILL, R. D. (1985). 'On estimating transition intensities of a Markov process with aggregate data of a certain type'. *Scandinavian Journal of Statistics*, 4.

—— and SCHUMACHER, M. (1987). 'A simple test of the proportional hazards assumption'. *Biometrika*, 74: 289–300.

GINSBERG, R. B. (1971). 'Semi-Markov processes and mobility'. *Journal of Mathematical Sociology*, 1: 233–62.

HECKMAN, J., and SINGER, B. (1982). 'Population heterogeneity in demographic models.' In *Multidimentional Mathematical Demography*, ed. R. C. Kenneth and A. Rogers. New York: Academic Press, pp. 567–604.

—— —— (1984). 'A method for minimizing the impact of distributional assumptions in econometric models for duration data'. *Econometrica*, 52: 271–320.

HENRY, L. (1959). 'D'un problème fondamental de l'analyse démographique'. *Population*, 13: 9–32.

—— (1966). 'Analyse et mesure des phénomènes démographiques par cohortes'. *Population*, 21: 465–82.

—— (1972). *Démographie: analyse et modèles*. Paris: Editions de l'INED, 1984.

HOBCRAFT, J., and MURPHY, M. (1986). 'Demographic event history analysis: a selective review'. *Population Index*, 52: 3–27.

HOEM, J. (1976). 'The statistical theory of demographic rates'. *Scandinavian Journal of Statistics*, 3: 169–85.

—— (1985). 'Weighting, misclassification and other issues in the analysis of survey samples of life histories'. In *Longitudinal Analysis of Labour Market Data*, ed. J. Heckman and B. Singer. Cambridge University Press.

—— (1987a). 'The issue of weights in panel surveys of individual behavior'. Research Report no. 39, Stockholm Department of Statistics.

—— (1987b). 'Statistical analysis of a multiplicative model and its application to the standardization of vital rates: a review'. *International Statistical Review*, 55: 119–52.

—— and FUNCK JENSEN, U. (1982). 'Multistate life table methodology: a probabilist critique'. In *Multidimensional Mathematical Demography*, ed.

K. Land and A. Rogers. New York: Academic Press, pp. 155–264.
HOGAN, D. P. (1978). 'Order of events in the life course'. *American Sociological Review*, 43: 573–86.
JAYET H. and MOREAU, A. (1988). 'Proportional hazard model: estimation and specification tests using asymptotic least squares'. Working Paper no. 8804, INSEE.
JOHANSEN, S. (1983). 'An extension of Cox's regression model'. *International Statistical Review*, 51: 165–74.
JOHNSON, N. L., and KOTZ, S. (1970). *Continuous Univariate Distributions*, 2 vols. New York: John Wiley.
KALBFLEISCH, J., and PRENTICE, R. (1973). 'Marginal likelihood based on Cox's regression and life model'. *Biometrika*, 60: 267–78.
—— —— (1980). *The Statistical Analysis of Failure Time Data*. New York: John Wiley.
KAPLAN, E., and MEIER, P. (1958). 'Nonparametric estimation from incomplete observations'. *Journal of the American Statistical Association*, 53: 457–81.
KAY, R. (1986). 'A Markov model for analysing cancer markers and disease states in survival studies'. *Biometrics*, 42: 855–65.
KEILMAN, N., KUIJSTEN, A., and VOSSEN, A. (1988). *Modelling Household Formation and Dissolution*. Oxford: Clarendon Press.
KEYFITZ, N. (1968). *An Introduction to the Mathematics of Demography*. Reading, Mass.: Addison-Wesley.
—— (1985). *Applied Mathematical Demography*. New York: John Wiley.
KIMBALL, T., and PEARSALL, M. (1954). 'The nature of human groups'. In *The Talladega Story*. University of Alabama Press.
KIRWAN, F. and HARRIGAN, F. (1986). 'Swedish–Finnish return migration, extent, timing, and information flows'. *Demography*, 23: 313–27.
KLIJZING, E., SIEGERS, J., KEILMAN, N., and GROOT, L. (1988). 'Static versus dynamic analysis of the interaction between female labour force participation and fertility'. *European Journal of Population*, 4: 97–116.
LAGAKOS, S., BARRAJ, L., and de GRUTTOLA, V. (1988). 'Nonparametric analysis of truncated survival data, with application to AIDS'. *Biometrika*, 75: 515–23.
LAND, K. C., and ROGERS, A. (ed.) (1982). *Multidimensional Mathematical Demography*. New York: Academic Press.
LANGEVIN, A. (1986). 'Rythmes sociaux et réinterprétation individuelle des critères d'âge dans le parcours de la vie'. *Annales de Vaucresson*, 25: 169–80.
LELIÈVRE, E. (1986). 'The analysis of interactions between phenomena: data (a French survey), tools, first results'. Paper presented to IIASA conference, Sopron, Hungary, September.
—— (1987a). 'Activité professionnelle et fécondité: les choix et les détermi-

nations des femmes françaises entre 1930 et 1960'. *Cahiers Québéquois de Démographie*, 16: 207–36.
LELIÈVRE, E. (1987*b*). 'Migration définitives vers la France et constitution de la famille'. *Revue Européenne des Migrations Internationales*, 3: 35–54.
—— (1988*a*). *Méthodes mathématiques et statistiques pour l'analyse d'histoires de vie*. Nouvelle Thèse de Doctorat de l'EHESS, March.
—— (1988*b*). 'Acquisition du premier logement et la naissance du dernier enfant'. In C. Bonvalet and A. M. Fribourg (eds.), *Congrès et colloques*, 2: 117–28.
—— (1988*c*). 'L'Étude des interactions entre phénomènes: dépendance unilatérale et causalité'. Paper presented to the AIDELF International Colloquium on 'Démographie et Différences', Montréal, 7–10 June.
—— (1988*d*). 'Constitution de la famille et urbanisation du Mexique'. In A. Quesnel and P. Vimant (eds.), *Migration, changements sociaux et développement*. Paris: Orstom.
—— and COURGEAU, D. (1991). 'Approches longitudinales'. In F. Stendler and P. Watier (eds.), *Interrogations et parcours sociologiques*. Paris: Meridiens Klincksieck.
LINDSAY, B. A. (1983*a*). 'The geometry of mixture likelihoods: a general theory'. *Annals of Statistics*, 11: 86–94.
—— (1983*b*). 'The geometry of mixture likelihoods: the exponential family'. *Annals of Statistics*, 11: 783–92.
LYBERG, I. (1983). 'The effect of sampling and nonresponse on estimates of transitions intensities: some empirical results from the 1981 Swedish Fertility Survey'. *Stockholm Research Reports in Demography*, no. 14.
MAU, J. (1986). 'On a graphical method for the detection of time-dependent effects of covariates in survival data'. *Applied Statistics*, 35: 245–55.
MAYER, K. U., and TUMA, N. B. (1987). 'Applications of event history analysis in life course research'. *Materialen aus der Bildungsforschung*, no. 30. Berlin: Max-Planck-Institute.
MCCULLAGH, P., and NELDER, J. A. (1983). *Generalized Linear Models*. London: Chapman & Hall.
MCGINNIS, R. (1968). 'Stochastic model of social mobility'. *American Sociological Review*, 33: 712–21.
MENKEN, J., and TRUSSEL, J. (1981). 'Proportional hazards life table models: an illustrative analysis of socio-demographic influences on marriage dissolution in the United States'. *Demography*, 18: 181–200.
MODELL, J., FURSTENBERG, F., and STRONG, D. (1978). 'The timing of marriage in the transition to adulthood: continuity and change, 1860–1975; turning points'. *American Journal of Sociology*, 84, Special Issue: 120–50.
MONNIER, A. (1987). 'Projets de fécondité et fécondité éffective, une enquête longitudinale: 1974, 1976, 1979'. *Population*, 42: 819–42.
MURPHY, M. (1984). 'The influence of fertility, early housing career and

socio-economic factors on tenure determination in contemporary Britain'. *Environment and Planning* A, 16: 1303–18.
MURPHY, M. and SULLIVAN, O. (1985). 'Housing tenure and family formation in contemporary Britain'. *European Sociological Review*, 1: 230–43.
NELSON, W. (1969). 'Hazard plotting for incomplete failure data'. *Journal of Qualitative. Technology*, 1: 27–52.
—— (1972). 'Theory and application of hazard plotting for censored failure data'. *Technometrics*, 14: 945–65.
NEVEU, J. (1986). 'Arbres et processus de Galton-Watson'. *Annales de l'Institut Henri Poincaré*, 22: 199–207.
OAKES, D. (1981). 'Survival times: aspects of partial likelihoods'. *International Statistical Review*, 49: 235–64.
PETERSEN, T. (1986a). 'Estimating fully parametric hazard rate models with time-dependent covariates'. *Sociological Methods and Research*, 14: 219–46.
—— (1986b). 'Fitting parametric survival models with time-dependent covariates'. *Applied Statistics*, 35: 281–8.
PETO, M., and PETO, R. (1972). 'Asymptotically efficient rank invariant test procedures (with discussion)'. *Journal of the Royal Statistical Society* A, 135: 185–206.
PETO, R., and PIKE, M. C. (1973). 'Conservatism of the approximation $\Sigma(O - E)^2/E$ in the logrank test for survival data or tumor incidence data'. *Biometrics*, 60: 579–84.
PICKLES, A., and DAVIES, R. B. (1983). 'The longitudinal analysis of housing careers'. *Journal of Regional Science*, 75: 85–101.
PONS, O., and DE TURCKHEIM, E. (1988). *Modèles de régression de Cox périodique et étude d'un comportement alimentaire*. Cahiers de Biométrie. Versailles: INRA.
POULAIN, M., RIANDEY, B., and FIRDION, J. M. (1991). 'Une expérimentation franco-belge sur la fiabilité des enquêtes rétrospectives: l'enquête 3B BIS', *Population*, 46: 65–88.
POURCHER, G. (1964). *Le peuplement de Paris: origine régionale, composition sociale, attitudes et motivations*. Cahier Travaux et Documents no. 43. Paris: INED/PUF.
PRENTICE, R. (1973). 'Exponential survivals with censoring and explanatory variables'. *Biometrika*, 62: 279–88.
—— (1975). 'Discrimination among some parametric models'. *Biometrika*, 62: 607–14.
—— (1982). 'Covariate measurement errors and parameter estimation in a failure time regression model'. *Biometrika*, 69: 331–42.
—— (1986). 'On the design of synthetic case-control studies'. *Biometrics*, 42: 301–10.
—— and BRESLOW, N. (1978). 'Retrospective studies and failure time models'. *Biometrika*, 65: 153–8.

PRESSAT, R. (1961). *L'Analyse démographique*. Paris: PUF.
—— (1966). *Principes d'analyse, cours d'analyse démographique de l'IDUP*. Paris: Editions de l'INED.
PRIGOGINE, I., and STENGERS, I. (1988). *Entre le temps et l'éternité*. Paris: Fayard.
RIANDEY, B. (1985). 'L'enquête biographie familiale, professionnelle et migratoire (INED 81): le bilan de la collecte'. In *Migrations internes*. Chaire Quetelet'83. Louvain-la-Neuve: Jezierzki.
RICOEUR, P. (1983). *Temps et récit*: i, *La configuration du temps dans le récit de fiction*; ii, *Le temps raconté*; iii, *L'ordre philosophique*. Paris: Seuil.
ROGERS, A. (1973a). 'The multiregional life table'. *Journal of Mathematical Sociology*, 3: 127–37.
—— (1973b). 'The mathematics of multiregional demographic growth'. *Environment and Planning*, 5: 3–29.
ROUY, E. (1986). 'Rapport de stage sur les troncatures à droite'. Unpublished paper.
SANDEFUR, G. (1985). 'Variation in interstate migration of men across the early stages of the life cycle'. *Demography*, 22: 353–66.
—— and SCOTT, W. (1981). 'A dynamic analysis of migration: an assessment of the effect of age, family and career variables'. *Demography*, 18: 355–68.
SCHOEN, R. (1979). 'Calculating increment–decrement life tables by estimating mean durations at transfer from observed rates'. *Mathematical Biosciences*, 47: 255–69.
SCHOU, G. and VAETH, M. (1980). 'A small sample study of occurrence/exposure rates for rare events'. *Scandinavian Actuarial Journal*, 4: 209–25.
SCHWEDER, T. (1970). 'Composable Markov processes'. *Journal of Applied Probability*, 7: 400–10.
SINGER, B., and SPILLERMAN, S. (1974). 'Social mobility models for heterogeneous population'. In *Sociological Methodology*, ed. Costner. San Francisco: Jossey-Bass, pp. 356–401.
SORENSEN, A. (1977). 'Estimating rates from retrospectives questions'. In *Sociological Methodology*, ed. D. R. Heise. San Francisco: Jossey-Bass, pp. 209–23.
—— WEINERT, F. E., and SHERROD, L. R. (ed.) (1986). *Human Development and the Life Course: Multidisciplinary Perspectives*. Hillsdale, NJ: Laurence Erlbaum Associates.
SPILLERMAN, S. (1972). 'Extension of the Mover–Stayer model'. *American Journal of Sociology*, 78: 599–626.
STRUTHERS, C., and KALBFLEISCH, J. (1986). 'Misspecified proportional hazard models'. *Biometrika*, 2: 363–9.
SUZUKI, K. (1985). 'Nonparametric estimation of lifetime-distributions from a record of failures and follow-ups'. *Journal of the American Statistical Association*, 80: 68–72.

SWAIN, M. (1987). 'Theories of causation'. In *L'explication en sciences sociales: la recherche des causes en démographie*. Chaire Quetelet '87 Louvain-la-Neuve: Ciaco, pp. 197–214.

TAPINOS, G. (1985). *Eléments de démographie*. Paris: Armand Colin.

TRUSSEL, J., and HAMMERSLOUGH, C. (1983). 'A hazard model analysis of the covariates of infant martality in Sri Lanka'. *Demography*, 20: 1–26.

—— and RICHARD, T. (1985). 'Correcting for unmeasured heterogeneity in hazard models using the Heckman–Singer procedure'. In *Sociological Methodology*, ed. N. Tuma. San Francisco: Jossey-Bass, pp. 242–76.

—— Hankinson, R., and Tilton, J. (eds.) (1992). *Demographic Applications of Event History Analysis*. Oxford University Press.

TSAI, W., LEURGANS, S., and CROWLEY, J. (1986). 'Nonparametric estimation of a bivariate survival function in presence of censoring'. *Annals of Statistics*, 14: 1351–65.

TUMA, N. (ed.) (1985). *Sociological Methodology*. San Francisco: Jossey-Bass.

—— and HANNAN, M. (1984). *Social Dynamics, Models and Methods*. Orlando, Fla.: Academic Press.

—— —— (1984). 'Models for heterogeneous populations'. In *Social Dynamics*. Orlando, Fla.: Academic Press, pp. 155–86.

—— —— and GROENEVELT, L. (1979). 'Dynamic analysis of event histories'. *American Journal of Sociology*, 84: 820–54.

TURNBULL, B. (1974). 'Nonparametric estimation of a survivorship function with doubly censored data'. *Journal of the American Statistical Association*, 69: 169–73.

VALLIN, J., and NIZARD, A. (1977). 'La mortalité par état matrimonial: mariage sélection ou mariage protection'. *Population*, Special Issue: 95–125.

VAUPEL, J., and YASHIN, A. (1985). 'Heterogeneity ruses: some surprising effects of selection on population dynamics'. Unpublished paper.

VINCENT, P. (1947). 'Nomogrammes pour la détermination des différences significatives entre deux taux'. *Population*, 2: 313–22.

WATERS, H. R. (1984). 'An approach to the study of multiple state models'. *Journal of the Institute of Actuaries*, 111: 363–74.

WENDEL, B. (1953). *A Migration Scheme: Theory and Observation*. Lund: Gleerup.

WING HUNG WONG (1986). 'Theory of partial likelihood'. *Annals of Statistics*, 14: 88–123.

WUNSCH, G. (1988). *Causal Theory and Causal Modeling*. Louvain: Leuven University Press.

YASHIN, A., MANTON, K., and VAUPEL, J. (1985). 'Mortality and aging in a heterogeneous population: a stochastic process model with observed and unobserved variables'. *Theoretical Population Biology*, 27: 154–75.

Subject Index

Aalen estimator 59–61, 73
accelerated failure time model
   141–3, 169–73, 180
  comparison with proportional
    hazards 142
  log-logistic 142–3, 169–73
actuarial estimator:
  of hazard function 71, 90

baseline hazard 180–1, 189
bias 51, 61, 67
birth 2, 10, 19, 21, 50, 96, 101–2,
   109, 196
  child 2, 29
  date of 21, 22
  first 26, 39, 62–5, 85, 102, 193,
    199
  last 19, 52, 87
  of the last child 85, 113, 151,
    153, 155, 159
  second 11, 85, 86, 199
  third 86, 193
bivariate case 82, 87–95, 99,
   101–2

censoring 50–67, 146–7, 181
  independant 147
  left 26, 50–1, 61–2
  right 11, 27, 52–3
  type of 50
cohabitation 94, 101
comparison of distributions 134–5,
   173–7
Competing risks 72–5
computer packages:
  EVACOV (INED) 189, 209
  GLIM 209
  LIFETEST (SAS) 78–80, 208
  LOGLIN 208
  PL2 (BMDP) 189, 209
  RATE 144, 209
  ROOT (INED) 95, 209

confidence interval 48, 92, 150
consensual union 10, 101
covariance matrix 54, 76, 91, 156,
   162, 164, 169, 188
cumulative intensity, *see* integrated
   cumulative hazard

death 2, 10–11, 20, 29, 50, 54, 56,
   82
  cause of 59, 72
  date of 21
density probability function:
  conditional, *see* hazard function
  joint 89
  non-parametric 30–1, 88
  parametric 110, 114, 118, 122,
    126, 128, 129, 131, 137, 138,
    142
dependance 83–7, 90
  a priori 86–7, 200
  recipocal 88, 199
  unilateral 84, 86, 88, 198
departure 57
  from the parents' home 27, 83,
    85, 93, 94
  from professional activity 96
  from the agricultural sector 199
divorce 94, 101
  date 9
duration of stay 6, 16, 19, 23–5, 27,
   29, 43, 52, 78–9, 106, 115, 123–
   4, 166–7, 173, 201, 204
dwelling 15, 174
  change of 16–17, 21, 24–5,
    115–16, 120, 123–4, 127
  date of arrival in the 19
  purchase of the first 191–2

educational level 135, 155, 157,
   162, 173
  status 109
emancipation 21–2

exponential distribution 110–13, 121, 126, 134, 137
   estimation of the parameter of 149–57, 174
   mixing 113–21

family 3, 4, 9, 189, 191
   history 12–14, 27
female activity 85, 87, 103–4
fertility 1, 9, 11, 37–8, 85, 87, 101–2, 104, 199–200
   hazard rate 97
   rate 20
   survey 20
Fisher information matrix 54, 148, 157, 170
Fisher–Snedecor distribution 133–4, 175
   estimation of the parameters of 175–6
fuzzy time 96–7

gamma distribution 128–9
Gompertz distribution 16, 121–5, 127, 134
   estimation of the parameters of 163–9, 174
Gompertz–Makeham distribution 124–5, 144
Greenwood formula 55–6, 69, 71

hazard function estimation 54, 68, 71, 90
   non-parametric 31–2, 46, 72, 87, 99, 104–5
   parametric 110, 114, 118, 121, 125, 128, 129, 131, 137, 138, 143
   semi-parametric 180
heterogeneity 2, 4, 30, 36, 109, 117, 135, 195, 197, 200–4
   unobserved 44–5, 178, 194, 202–3
home owner 113, 132–2, 139–40, 151, 153, 155, 159, 162–3, 174, 191

incomplete observation 18–26

independance 84, 88–9, 198
instantaneous hazard rate, *see* hazard function
instantaneous rate of failure, *see* hazard function
interaction 1, 3, 36, 83, 85–6, 93–4, 95, 101, 104, 137, 189, 203
   analysis 9–10, 99

Job:
   at marriage 85
   change 29, 52, 102–3
   first 27, 29, 47, 78–80, 83, 94
   in farming 84, 189
   mobility 109, 114

Kaplan–Meier estimator 53, 55, 68, 73, 186
likelihood:
   function 32–7, 47, 54, 145–7, 157
   log 69, 148, 150
   partial 32
log-logistic distribution 131–3, 134–5, 138, 143
   estimation of the parameters of 169–73, 176
   with accelerated failure times 142–3
log-normal distribution 129–31, 176

Markov 30, 38–45
marriage 2, 4, 10–11, 14, 21, 24–5, 27, 29, 38, 46, 84–5, 87, 94–5, 101, 109, 189, 191, 193, 199
   age at 5, 26, 105
   date of 9, 21–2, 29, 61
   duration 17, 23
   first 50
   status 9
martingale 60
maximum likelihood estimator 47, 54, 70–1, 146, 150, 152, 157
memory 12
   errors 20–6
migration 1–2, 26, 29, 38–9, 43, 46, 50, 54, 61, 68, 101, 104, 109, 136, 173–4, 199

date of 9, 21
  first 26, 65–6, 94, 102
  hazard rate of 77
  history 12–13, 27
  in- 21, 95
  internal 10, 114
  last 19, 52
  multiple 9
  out- 12, 21, 77
  prenuptial 15
  rate 16, 23
  to metropolitan areas 85, 95, 199
  to non-metropolitan areas 95
mixing distribution:
  exponential 113–21
mortality 1, 9, 18, 37, 59, 74–5
  cumulative hazard curves of 75
  rate 18
move 1, 29
  previous 26
mover-stayer model 45, 114, 117–120, 123, 127

Nelson:
  estimator 60, 73
  plot 73
Newton–Raphson algorithm 157–8, 161–2, 165, 170, 176, 187–8
non-multiplicative hazard model 202
nuptiality 1, 9, 12, 37–8, 87, 95, 189, 204
  hazard rate of 95, 190
  rate 18, 37

occupation 15
  change 29, 38, 46
  last 19
  professionnal 43
  status 10, 15, 174
occupational status 78–81
order statistic 182

Pareto distribution 118–19
Poisson 41–2, 52–3
population register 1–2, 10
product integral 34
product–limit estimator, *see* Kaplan–Meier estimator

professional career 11, 78
  life 93, 173
proportional hazard model 136–43, 165, 180–1, 188, 202

random loss 146
rank test 75–6, 91, 182
residence 43
  change of 19
  place of 9, 14
  region of 39

sampling plan 17
  informative 18–19
  non-informative 14, 17–18
separation 101–2
simultaneity 90, 95–7, 100
spatial mobility 24–5
survey 11
  multiround 11
  prospective 11, 28
  retrospective 11–13, 26–8
  'Triple biographie' 13–17, 20, 48, 61, 63, 77, 113, 115, 151, 162, 166, 173, 189
survivor function:
  estimation 55, 68, 71, 185–7
  non-parametric 2, 30, 88
  parametric 110, 114, 118, 121, 126, 128, 129, 131, 137, 138, 142
  semi-parametric 185

time 4–5, 10, 29–30, 37, 43–8
  continuous 30, 33–5
  dependant characteristics 178
  discrete 32, 33–5
  interval 41, 50
  of interview 10
  of occurence 46
  at risk 150
  of survey 28
  space- 3
  waiting 52

variance 54, 76, 145, 149, 150, 152
  asymptotic 55–6, 68–9, 71

Weibull distribution 125–7, 134, 138–9
  estimation of the parameters of 158–63, 76
weighting 16–18

work:
  history 12–16, 27
  place of 3
  start to 199

Author Index

Aalen, O. 59, 74, 91
Allison, P. 36
Arjas, E. 36

Blumen, I. 114
Bretagnolle, J. 203

Coale, A. 45
Courgeau, D. 15, 16, 21, 23, 26, 39, 43, 61, 84–5, 87, 96, 100, 104, 114, 166, 173, 189, 191, 199, 201–2, 204
Cox, D. 90, 148–9
Crowley, J. 188

Deroo, M. 20
Duchêne, J. 10, 21
Dussaix, A. 20

Elder, G. 28

Feller, W. 42, 51
Firdion, J. 10, 12, 15, 21–2
Foner, A. 28
Funck Jensen, U. 40

Ginsberg, R. 43
Groot, L. 96–7

Heckman, J. 178, 203
Henry, L. 1–2, 37, 195
Hinkley, D. 148–9
Hoem, J. 18–19, 40, 91
Hu, M. 188
Huber-Carol, C. 203

Johnson, N. 129

Kalbfleish, J. 47, 110, 134, 147, 182
Kangas, P. 36
Kaplan, E. 27, 53, 58

Keilman, N. 96–7
Keyfitz, N. 38
Kertzer, D. 28
Kimball, S. 3
Klijzing, E. 96–7
Kotz, S. 129

Langevin, A. 28
Lelièvre, E. 16, 84–5, 87, 96, 100, 104, 189, 191, 199, 201–2, 204
Lyberg, I 12, 20

McCarthy, P. 114
McGinnis, R. 43
McNeil, D. 45
Marvin, K. 114
Meyer, P. 27, 53, 59
Modell, J. 28
Monnier, A. 11
Murphy, M. 193

Nelson, W. 59
Nizard, A. 18

Oakes, D. 90

Peto, R. 91
Pike, M. 91
Poulain, M. 10, 12, 15, 21–2
Pourcher, G. 13
Prentice, R. 47, 110, 134, 147, 176, 182
Pressat, R. 1, 195
Prigogine, I. 4

Riandey, B. 10, 12, 14–15, 21–2
Richards, T. 178, 203
Rogers, A. 36
Rouy, E. 63

Schow, G. 92
Siegers, J. 96–7
Singer, B. 44, 178, 203

Spillerman, S. 44
Stengers, I. 5

Trussel, J. 178, 203
Turnbull, B. 62

Vaeth, M. 92
Vallin, J. 18

Wendel, B. 10